GREAT DISCOVERIES IN SCIENCE
Plate Tectonics

Fiona Young-Brown

Published in 2019 by Cavendish Square Publishing, LLC
243 5th Avenue, Suite 136, New York, NY 10016

Copyright © 2019 by Cavendish Square Publishing, LLC

First Edition

No part of this publication may be reproduced, stored in a retrieval system, or transmitted in any form or by any means—electronic, mechanical, photocopying, recording, or otherwise—without the prior permission of the copyright owner. Request for permission should be addressed to Permissions, Cavendish Square Publishing, 243 5th Avenue, Suite 136, New York, NY 10016. Tel (877) 980-4450; fax (877) 980-4454.

Website: cavendishsq.com

This publication represents the opinions and views of the author based on his or her personal experience, knowledge, and research. The information in this book serves as a general guide only. The author and publisher have used their best efforts in preparing this book and disclaim liability rising directly or indirectly from the use and application of this book.

All websites were available and accurate when this book was sent to press.

Library of Congress Cataloging-in-Publication Data

Names: Young-Brown, Fiona, author.
Title: Plate tectonics / Fiona Young-Brown.
Description: First edition. | New York : Cavendish Square, [2018] | Series: Great discoveries in science | Audience: Grades 9 to 12. | Includes bibliographical references and index.
Identifiers: LCCN 2018013139 (print) | LCCN 2018014490 (ebook) | ISBN 9781502643742 (ebook) | ISBN 9781502643841 (library bound) | ISBN 9781502643964 (pbk.)
Subjects: LCSH: Plate tectonics--Juvenile literature. | Geology, Structural--Juvenile literature. | Earth sciences--History--Juvenile literature.
Classification: LCC QE511.4 (ebook) | LCC QE511.4 .Y67 2018 (print) | DDC 551.1/36--dc23
LC record available at https://lccn.loc.gov/2018013139

Editorial Director: David McNamara
Editor: Jodyanne Benson
Copy Editor: Michele Suchomel-Casey
Associate Art Director: Alan Sliwinski
Designer: Christina Shults
Production Coordinator: Karol Szymczuk
Photo Research: J8 Media

The photographs in this book are used by permission and through the courtesy of: Cover Semnic/Shutterstock.com; p. 4 Fotos593/Shutterstock.com; p. 9 Lynette Cook/Science Photo Library/Getty Images; p. 10 Ian Cuming/Ikon Images/Getty Images; p. 12 Vadim Sadovski/Shutterstock.com; p. 16 D'Arco Editori/De Agostini Picture Library/Getty Images; p. 17 Lara-sh/Shutterstock.com; p. 19 Dorling Kindersley/Getty Images; p. 22 Paul D Stewart/Science Photo Library/Getty Images; p. 26 John Crux/Shutterstock.com; p. 28 Jakinnboaz/Shutterstock.com; p. 31 G Brad Lewis/Science Faction/Getty Images; p. 34 William Frederick Mitchell/Wikimedia Commons/File:HMS challenger William Frederick Mitchell.jpg/CC0; p. 37 Unknown/Wikimedia Commons/File:PSM V29 D322 Gray and milne seismograph with recorder.jpg/CC0; p. 40 Eduards Normaals/Shutterstock.com; p. 47 Tinkivinki/Shutterstock.com; p. 48 Print Collector/Hulton Archive/Getty Images; p. 51 Evikka/Shutterstock.com; p. 54 National Science Foundation/Wikimedia Commons/File:Morgan, W. Jason.jpg/CC0; p. 59 Charles O'Rear/Corbis Documentary/Getty Images; p. 62 Fritz Goro/The LIFE Picture Collection/Getty Images; p. 67 SPL/Science Source; p. 70 Robin2/Shutterstock.com;p. 75 LouieLea/Shutterstock.com; p. 80 Designua/Shutterstock.com; p. 82 Vitoriano Junior/Shutterstock.com; p. 87 ONYXprj/Shutterstock.com; p. 88 USGS/Wikimedia Commons/ File:Bathymetry image of the Hawaiian archipelago.png/CC0; p. 92 Ventdusud/Shutterstock.com; p. 94 Everett Historical/Shutterstock.com; p. 101 Janez Volmajer/Shutterstock.com; p. 104 Mike Korostelev https://www.mkorostelev.com/Moment/Getty Images; p. 107 Iurii/Shutterstock.com.

Printed in the United States of America

Contents

Introduction	5
Chapter 1: The Problem of Understanding the Earth	11
Chapter 2: The Science of Plate Tectonics	29
Chapter 3: The Major Players in Plate Tectonics	49
Chapter 4: The Discovery of Plate Tectonics	71
Chapter 5: The Influence of Plate Tectonics	93
Chronology	110
Glossary	114
Further Information	118
Bibliography	120
Index	124
About the Author	128

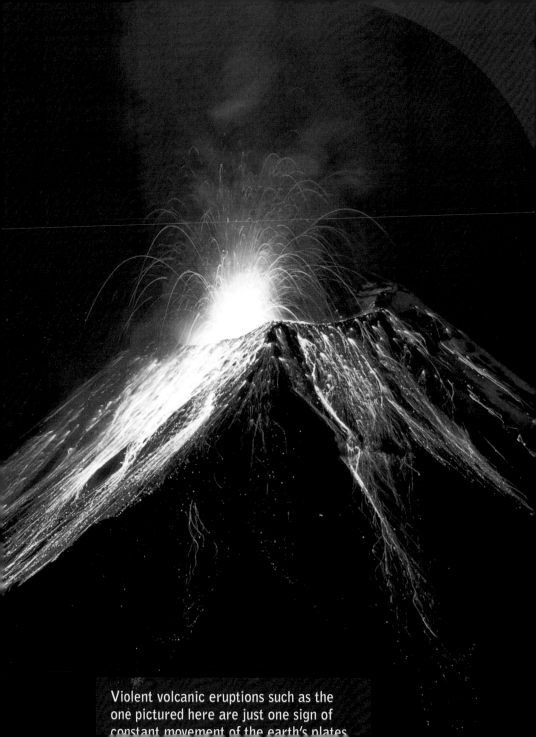

Violent volcanic eruptions such as the one pictured here are just one sign of constant movement of the earth's plates

Introduction

One afternoon in February 2018, the rush hour traffic of Mexico City came to an abrupt halt as the ground beneath started shaking violently. People ran for cover as buildings swayed in the 7.2 magnitude earthquake. Fortunately, the damage was minimal and there were no deaths. For some older residents, it was a frightening reminder of the devasting 8.0 earthquake that hit the city in 1985. That time, they were not so lucky. The city suffered extensive damage, and more than five thousand people lost their lives.

As Mexico City trembled, thousands of miles away, the red-hot lava continued to bubble in Ethiopia. The volcanic Erta Ale, one of the world's few lava lakes, has been erupting regularly since the 1960s.

Going back a few years, we can visit Honshu, Japan. This island nation is no stranger to natural disasters, but that does not make the widespread damage any less tragic. In March 2011, a powerful 9.0 earthquake off the coast of Tohoku was followed by massive tsunami waves that stood more than 128 feet (39 meters) high and traveled as far as

6 miles (10 kilometers) inland. The waves damaged a nuclear power plant, and people for miles around were forced to flee their homes. Many still have not been able to return. The total death toll was more than fifteen thousand with many more injured or missing.

Volcanic lava in Africa. Earthquakes in Mexico. Tsunamis in Japan. What links these fascinating but often catastrophic events?

The answer is plate tectonics.

It is only fairly recently that we discovered the answer. Not until the 1960s did scientists make a breakthrough that would revolutionize geological studies.

If we travel far back in history, we find countless myths and legends of the earth opening up, of angry gods sending plumes of fire into the sky, of villages being laid waste. The story of the destruction of Pompeii in 79 CE has been told worldwide. Now visitors flock to see the remains in Italy, which offer a snapshot of daily life many centuries ago.

The ancient Romans and Greeks attempted to explain why such things happened as they tried to make sense of nature. The Japanese engraved wood carvings and told tales of the angry fish that caused the waves to engulf the land. For those who lived in areas prone to natural disasters, rebuilding was a way of life.

The Renaissance saw a renewed interest in science. Artists such as Leonardo da Vinci collected fossils and wondered how sea creatures could have lived in the mountains. Others drew the first detailed atlases and marveled at how it seemed that different countries could almost slot together.

By the eighteenth century, the fields of geology and paleontology were developing, but they were regarded as hobbies rather than proper science. Finally, the Scotsman James Hutton published a controversial theory that the earth was millions of years old and that the rocks that make up the planet's crust were constantly changing as a result of heat, pressure, and erosion.

The Industrial Revolution of the nineteenth century triggered new scientific curiosities and advances. Charles Darwin investigated the origins of man and animals. Ocean floors were charted for the first time, and mountains were found deep below the seas. Each new discovery raised new questions.

Then, in 1912, a meteorologist from Germany made a bold claim. Alfred Wegener argued that all of the continents were once joined together. This, in itself, was not particularly new. Others had suspected as much. However, they had all believed that the continents had been joined by land bridges, now collapsed beneath the oceans. Wegener was suggesting something entirely different: they had all been part of one large mass before they drifted apart.

His theory of continental drift was openly ridiculed by many. But over the years, researchers found themselves frequently circling back to the notion. Each new discovery seemed to suggest that there might be something to it, after all. If only they could explain how such a thing might have occurred.

The answer lay in plate tectonics.

In this book, we will take a look at the field of plate tectonics and explore how it helped us to understand the natural occurrences of our planet. First, we will look at how the ancient Greeks studied what was happening around them. We will learn how they viewed the earth as flat and how myths provided an explanation for the ground that shook and the fire that flew into the sky. Then, we will look at how fledgling scientists of the Renaissance explored the world around them. As they made careful records of the fossils they found, they envisaged a world once covered by water. But, how did living creatures come to be trapped inside of solid rock?

In chapter two, we will chart the progress made in geology, oceanography, and seismology. As researchers advanced, they found that these separate subjects were all more closely related than expected. Chapter three will take a closer look at some of the people making the discoveries. There are biographies of Wegener and the men who discovered plate tectonics as well as the stories of those who filled in the gaps with their studies of earthquakes and more.

Chapter four looks at the events leading up to the discovery of plate tectonics and the discovery itself. We then explore what we know about the continental plates and how their interactions cause a variety of seismic events, some of which forever alter our landscape.

In chapter five, we look forward. We look at how our expanded knowledge of plate tectonics has helped us to improve safety in earthquake-prone areas through better construction methods. We explore how some buildings can withstand such violent shocks. Then, we turn our attention to space exploration. Probes have discovered the presence of volcanoes in faraway moons and evidence that plate tectonics may once have shaped the surface of Mars. What

This artwork depicts the tsunami caused by the explosion of the Indonesian volcano Krakatoa on August 27, 1883.

is there to learn from these findings? Finally, we dive deep into the future. How will plate tectonics continue to shape the earth? What could we expect to see if we return in 250 million years?

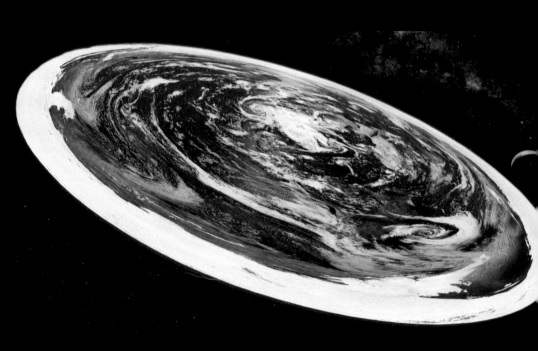

The belief that the earth is a flat disk has persisted through history, despite scientific evidence to the contrary.

CHAPTER 1

The Problem of Understanding the Earth

From some of the oldest myths and legends, we know that early man believed the earth was flat. It was a round disk that floated on air. At first, many explanations centered on the whims of the gods. Some peoples believed that gods lived within volcanoes. For example, a Hawaiian myth tells of Pele, the fire goddess, whose home is in a volcanic crater. As man's knowledge and development advanced, philosophers put forth various theories to try to explain natural phenomena such as weather, earthquakes, and so on. Gradually, scientific ideas also arose.

The Greek philosopher Anaxagoras suggested, some 2,500 years ago in 500 BCE, that everything in the universe was once made of one giant mass. An unknown but powerful force then made everything spin very fast. This caused the mass to break apart. The process created the planets, the stars, and the sun. The earth was at the center of the universe. It was heavy, made of iron, and could not move or rotate. Fire was lighter and so that made the stars and the sun, which all moved around the earth.

Anaxagoras went on to explain other phenomena. The earth floated on air. Sometimes tiny cracks would form deep within the surface, and air would rush up through the cracks. The air moved so quickly that it made the ground shake violently. Thus, he explained how earthquakes happened.

Centuries later, the Vikings would continue to believe that the earth was a flat disk and that if they sailed too close to the edges, they would fall off. (The edges, according to

The ancient Greeks tried to explain the movement of the sun and the earth as the reason for the seasons. But which was at the center of everything?

Norse legend, were protected by a terrifying sea dragon. You would be eaten by the dragon before you reached the edge of the earth.) Some Native American myths also say that the world is flat and that it lays on the back of a turtle. Those indigenous peoples along the rift lines in Northern California would explain earthquakes as what happened when the trickster Coyote tugged on a rope to make the ground move.

Meanwhile, back in ancient Greece, approximately 150 years after Anaxagoras attempted to explain the universe, Aristotle put forth the suggestion that the earth was round. Everything else—the stars, sun, moon, and other planets—revolved around it. This was nothing new since people already believed the earth was round—a round but flat disk. However, Aristotle couldn't help but notice that the stars in the north were not the same ones that he saw in the south. He reasoned that if the earth was flat, he should surely be able to see all of the stars whenever he looked into the sky. Likewise, the mountains in the distance should always be visible in their entirety. Therefore, he concluded that the earth must be a sphere instead of a disk. Aristotle also found some fossilized fish on land and theorized that they had once been looking for food and had become stranded. This encouraged people to think about how the land and the oceans may have changed or moved over time.

In approximately 200 BCE, another Greek scholar, the mathematician and astronomer Eratosthenes, became the first person to try to calculate the size of the earth. By measuring a series of angles between the land and the sun from two different locations, he estimated the earth to be roughly 22,798 miles (36,689.8 km) in circumference.

Myths to Explain the Earth

Around the globe, people have long used myths as a way of explaining natural events in ways that they can understand. These myths predate science but offer insight into a culture's history.

Many myths used stories based on local wildlife to help explain earthquakes. In India, it was said that elephants held up the earth. When an earthquake happened, it meant that one of the elephants was getting tired.

In China, it was a giant frog that held the earth; in Japan, a catfish. When either moved, it caused a tremor.

In parts of east Africa, legend says that there is a fish with a rock on it. On top of the rock is a cow. The earth balances on one of the cow's horns. Occasionally, the cow tosses the earth to the other horn, creating an earthquake.

Myths are also used to explain volcanic eruptions. In Hawaii, the goddess Pele can cause earthquakes simply by stamping her feet. When she erupts in anger, she causes a volcano to erupt.

Early Christians in northern Europe believed that the Icelandic volcano Hekla was a doorway to hell when it began to erupt.

As fiery magma flew from the crater, they believed that spirits were trying to escape the underworld. The noise of the eruption was caused by the screaming souls of the dead.

There is a developing field of study called geomythology. It is also referred to as "legends of the earth," "myths of observation," and "natural knowledge." Geomythology is the study of oral traditions created by pre-scientific cultures to explain (usually through poetic metaphor and mythological imagery) geological phenomena such as volcanoes, earthquakes, floods, and fossils. One type of geomyth includes folk explanations of significant geological features.

More researchers are starting to pay attention to these types of myths, believing that they may be helpful in teaching us about a region's geology.

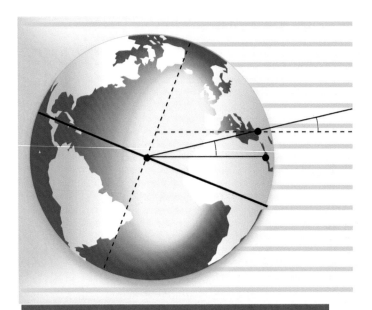

Eratosthenes used mathematical equations to calculate the circumference of the earth.

His measurements were surprisingly accurate. Modern scientists say that the circumference is 24,901 miles (40,074.3 km).

Anaxagoras had attempted to explain how and why earthquakes happen, but for the most part, Greek scholars were focused on the larger universe rather than what was taking place beneath our feet. Not until 20 CE did a theory arise as to why volcanoes erupted.

A man called Strabo wrote the first known geography book. In it, he detailed his many travels from Egypt, Sudan, and Ethiopia to Italy and the Mediterranean. This incredibly detailed work provided insight into the lands and peoples of the known world. At some point during his travels, he witnessed a volcanic eruption. He set about trying to explain the occurrence. Like Anaxagoras long before him, Strabo believed that hot air was the cause of earthquakes. Whereas the former had believed the air pushed its way through the disk that was the earth, Strabo knew that the earth was round. Therefore, he believed that pockets of hot air were trapped within the sphere. They would force their way to the surface, erupting in a mass of ash, steam, and dirt. These same underground air pockets caused earthquakes. Earthquakes and volcanic eruptions seemed to be connected.

POMPEII

Just fifty years or so after Strabo published his ideas, the world experienced one of the most famous volcanic eruptions in history. The Italian towns of Pompeii and

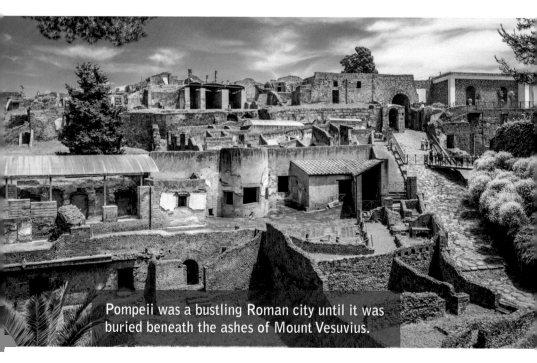

Pompeii was a bustling Roman city until it was buried beneath the ashes of Mount Vesuvius.

The Problem of Understanding the Earth   17

Herculaneum were, like their neighbors, used to the ground shaking. Earthquakes were an accepted part of life, and archaeologists have found evidence in Pompeii of damage from previous tremors, followed by repairs and rebuilding. The tremors they experienced during July and August 79 CE were no different.

Of course, we now know that a fault line runs the length of Italy, explaining the frequent tremors. But there was something else. At the time, the Romans did not understand, as we do today, how earthquakes and volcanic activity are connected. So when nearby Mount Vesuvius started to send up a column of smoke on the morning of August 24, bystanders watched with "a response more of curiosity than alarm" according to historian Pliny the Younger. Today, when a volcano starts showing signs of activity, scientists quickly monitor the situation and evacuate as necessary. But it seems that the residents of Pompeii and Herculaneum watched and carried on with their daily tasks. The plume of smoke continued throughout the day, but by evening, things were becoming more serious, and some people had already started to make their way to the coast to seek shelter.

Close to midnight, the volcano erupted in a vivid display of fire and lava. This red-hot pyroclastic flow burst forth, trapping anyone in its path. It quickly engulfed both Pompeii and Herculaneum. People had no time to escape. Today, we know much more about seismic activity and can hopefully avoid another occurrence as serious as this one. However, Pompeii remains a tourist attraction for many since so much of the town and its residents have been carefully preserved for centuries in the cooled lava.

Tsunamis are another result of plate movement and can cause massive damage to coastal and island communities.

Meanwhile, halfway around the globe, the island nation of Japan would have its own questions about why the earth shook. Japan is located at the junction of four tectonic plates. So, it has a long and very active seismic history. There are written records of some of the major seismic events, but, for the most part, people were more interested in rebuilding. Just as the residents of Pompeii seem to have become used to the fairly regular occurrence of earthquakes, so the people of Japan saw it as a part of everyday life, even if they could not explain it. There was one additional question raised by earthquakes in Japan though. Often after an earthquake, they would experience another natural phenomenon. The oceans would seem to rear back and then come crashing across the land in a wall of water. These giant waves were called tsunamis. Were tsunamis somehow related to earthquakes? If so, how? Were the forces that could make the ground shake also capable of controlling the oceans? No doubt these questions troubled Japanese philosophers just as the Greeks and Romans pondered their experiences with the natural world.

REVIVAL of INTEREST in the RENAISSANCE

After the decline of the Greek Empire and the Roman Empire, scientific developments slowed. Scholars continued to believe that the earth was at the center of the universe for more than a thousand years after Aristotle had lived. Not until the sixteenth century and the Renaissance did scientific exploration again take important strides, contributing to what we know today about the planet.

In the 1500s, Nicolaus Copernicus wrote that, contrary to prevailing beliefs, the earth was not at the center of the universe. Instead, the sun was in the center, with the earth and other planets circling it. At the time, he was punished by the church for saying such things. Yet his scholarship changed man's understanding of the universe and our own planet.

Meanwhile, someone else was laying the groundwork for later geological findings. Today, Leonardo da Vinci is known for many other achievements—painting, sculpture, engineering, to name but a few. He is less remembered for his studies of fossils, studies that would lead some to call him the father of paleontology and ichnology.

Leonardo was born in Tuscany, Italy, in 1452. There are very few records to tell us anything about his childhood, except that his education seems to have been quite informal but, as would be the case his entire life, driven by a fierce curiosity to understand how things work. When he was fourteen, he became an apprentice to a famous painter and sculptor. For the next seven years, he learned not just art, but all of the technical skills that came with it (metalwork, chemistry, and mechanics). After qualifying as a master guildsman, Leonardo worked on artistic commissions for various churches, and he also served as a military engineer. His skill range was broad, and he is considered one of the greatest minds of his time. When he died in 1519, it is said that the king of France held him in his arms and wept.

One of Leonardo's many interests was the study of fossils. As an artist, he paid keen attention to everything around him, including the landscape and rocks. He would often wander near his home gathering and examining the fossils that he found in the Italian countryside. At the time, people believed that fossils were some sort of mistake of creation. The rocks were trying to produce life. This explained why you might find a piece of bone, a shell fragment, or what looked like a tooth embedded in a piece of rock so far from the sea. Leonardo reached a difference conclusion. He kept a private diary where he sketched ideas and recorded his many findings. Here he wrote of his belief that the fossils he found were once living creatures or parts of creatures. How else could one explain the signs that insects had worked their way into the rocks? The creatures must have once lived in an ancient sea that had disappeared

Leonardo da Vinci was not just a talented artist and inventor. He was also fascinated by fossils and their origins.

long ago. As the shells and fragments dried out, insects had crawled into the mud that surrounded them. They then lay buried in centuries of rock and dirt.

Fossils provided man with some of the earliest clues about the movement of land and the oceans. But what is a fossil? How is it formed? Fossils can be any type of preserved remains from an earlier stage in geology. Leonardo found the fossils of shark teeth and shells, but fossilized items can also include piece of bone, wood, footprints, or even the detailed shape of a creature that has decayed but left an imprint in the rock around it.

One of the most common ways a fossil is formed is when the artifact is in a watery or muddy location. As it decays, it is covered in layers of mud and silt. Over thousands of years, as more soil builds up on top, the pressure causes much of the soil to harden. It becomes sedimentary rock. The soft tissues have rotted away, which is why some creatures such as jellyfish leave no fossil record. Meanwhile, the bones or shells are trapped inside the rock. The bones may decay, leaving a detailed imprint. Minerals could also leach into them from the surrounding soil, causing a process known as petrification. The petrified object turns into a type of stone.

Other items that have been found in fossilized form include dinosaur eggs, nests, dung, and footprints. Insects may become trapped in tree sap. Over time, this hardens into amber with the fossilized bug inside. It is also possible that Leonardo may have come across some fossils caused by volcanic eruptions. A creature becomes trapped in the volcanic ash or lava. The heat will destroy the tissue, but as the ash hardens into rock, the shape is preserved inside.

Since fossils need a wet environment to form, many of those we find are of sea creatures. However, many fossilized dinosaur remains have been found around the world, adding to our knowledge of early life. Fossil discoveries in areas hundreds of miles from the sea also tell us about the history of the earth. Huge numbers of fossils have been found in parts of Nebraska and Mongolia, telling us that these vast plains may once have been vast oceans. As the next few chapters will explain, plate tectonics means that it is entirely possible for the fossils of sea creatures to be found at the top of a mountain range.

Sadly, Leonardo never published his theories about where fossils came from. Not until nearly two hundred years later did another scientist start to study fossils. His research, uninfluenced by Leonardo's, added to the growing field of geology.

Nicolaus Steno was born in Copenhagen, Denmark, in 1638. He trained in the Netherlands as an anatomist and later became the personal physician of Grand Duke Ferdinand II in Florence. In addition to his advanced medical knowledge (he discovered the saliva ducts), Steno had a keen interest in geology. Much like Leonardo, he collected fossils as he traveled around Italy and kept detailed notes and sketches. Unlike Leonardo, Steno published his findings.

In 1669, his book, which carried the rather lengthy title *The Prodromus of Nicolaus Steno's Dissertation Concerning a Solid Body Enclosed by Process of Nature Within a Solid*, became one of geology's key early texts. In it, the author made several important claims. For starters, he believed that fossils had once been living creatures. Since living creatures could not possibly have made their way inside a piece of rock, the rock must have formed around them. He concluded that the rock was once a softer material, most likely a fluid that could flow over and through other materials. It then set and became hard.

Another important geological discovery by Steno was his theory that there were layers of history within the earth's crust. If we studied the different layers in the ground, we could learn about the past. Lastly, he believed that the crust had somehow moved and that this explained the formation of mountains. At the time, many people thought that mountains were living things, growing out of the ground like trees. Steno understood that they were actually formed by some kind of change on the earth's crust.

Like many of his contemporaries who made important theories about the universe, Steno's findings were considered controversial by the church. At the time, religious leaders stated that the earth was no more than six thousand years old. How could all of the changes that Steno mentioned take place within such a brief time span? He would never discover the answers. At the age of twenty-nine, Steno converted to Catholicism. He abandoned science entirely, instead becoming a priest, and later a bishop, for the remainder of his life.

Steno's theories may seem elementary now, but in the seventeenth century, they were major leaps forward in geological terms. He had suggested a completely new way of thinking about the earth's crust and the changes it had undergone with time. From his work, we have several geological principles. The principle of original horizontality says that rock and sediment all

Guatemala's Volcán de Fuego ("Volcano of Fire") is one of Central America's most active volcanoes.

create horizontal layers on top of each other. The law of superposition takes this further. It states that if there are layers of rock, the oldest layers must be those at the bottom. Newer layers form on top.

The discoveries made by Nicolaus Steno are very similar to those written about by Leonardo da Vinci in his journal. If Leonardo's journal had been published, geological science may have developed much more quickly. Likewise, if Steno had not abandoned his scientific career, who knows what further discoveries he might have made. Nevertheless, he provided inspiration for many others who would follow in his footsteps.

As we shall read in the next chapter, they added to our knowledge of the earth, its crust, the layers beneath, and how they all interact with each other. Their research would eventually help us comprehend what happens to cause a volcanic eruption, such as that which laid waste to Pompeii so many centuries earlier.

4 Layers of the Earth

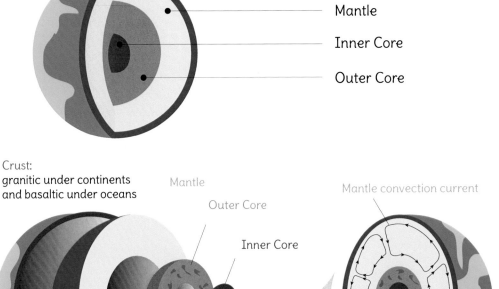

The earth is made up of four different layers, as shown in this picture.

CHAPTER 2
The Science of Plate Tectonics

We now know a great deal about the earth and how it is made up. We know that there are four layers: an inner and outer core, a mantle, and the outer crust. We also know that the crust is made up of plates that are in constant motion. Much of our understanding of plate tectonics is relatively recent. New key breakthroughs were not made until the 1960s, little more than fifty years ago.

On the other hand, it was the work of scientists and scholars in the Renaissance who helped to shape our understandings. They laid the groundwork for future researchers to build upon. In the last chapter, we read how Leonardo da Vinci made important observations that were never shared publicly. Then, Nicolaus Steno made the same observations in the 1600s. His findings advanced the interest in rocks and how they were formed. Steno gave us the principle of original horizontality and the law of superposition, telling us that layers of rock create horizontal beds, with the oldest layers at the bottom.

DIFFERENT TYPES of ROCK

In the early eighteenth century, scientists began to understand more about rocks with the realization that there were, in fact, two separate types of rock. Anton Moro was an Italian monk with an interest in rocks and nature. Like others before him who had studied in the Italian countryside, he found fossils of sea creatures high up in the mountains. This led him to conclude that the entire area must have once been under water.

Moro also spent a great deal of time exploring the small volcanic islands near the coast of Italy. As he did so, he noticed that the rocks there were different from the rocks he had found on the mainland. They seemed to be composed of different materials, and he became convinced that they were formed by different processes. Much of the rock he had found in the mountains and in earlier explorations was layered. This matched Steno's observations about layers of sediment and dirt forming one on top of the other, the bottom layers becoming more compressed by the pressure. This type of rock is what we call sedimentary.

But the rock on the volcanic islands was not like this at all. Moro called this volcanic rock. He proposed that they were rocks that had melted deep inside the earth before being thrown out during volcanic eruptions. Once the liquid rock reached the surface, it again cooled and hardened. Moro's volcanic rock is what we call igneous.

Volcanic lava was once an intensely hot, molten rock beneath the earth's crust. Once it has burst onto the earth's surface, it eventually cools, hardening into rock.

The BIRTH of MODERN GEOLOGY

At this point in history, the study of rocks was an interesting pastime, and certainly there had been some important observations made. However, geology was still not considered a science. That was about to change.

James Hutton was born in Edinburgh, Scotland, in 1726. The son of a merchant, he received a good education and became an apprentice to a lawyer. He quickly realized that a career in law was not for him and instead studied chemistry and medicine in Edinburgh, Paris, and the Netherlands. He spent a little bit of time in Edinburgh working in chemistry, but then he inherited some farmland. He spent the next few decades as a gentleman farmer, studying agriculture and seeking ways to improve farming methods. One might wonder how this led to rocks. Hutton, as both a farmer and a chemist, took an interest in the land. He noticed how the lands he farmed were affected by external forces such as the weather. Traveling around Britain, Hutton took note of the types of fossils he found and the different geological features he encountered. In 1768, he returned to Edinburgh to write about his observations and further develop his ideas.

Like Moro, Hutton observed that there were different types of rock and that the earth seemed to be undergoing a constant degree of change. Rocks formed, broke down, and changed into other forms. All of this, he reasoned, must have taken an incredible amount of time. It must have also needed a huge amount of force, such as heat, to cause solid rock to melt.

In the 1700s, science was still in its early stages of growth and religious beliefs continued to shape many theories. For example, it was widely accepted that the earth was about six thousand years old and that landforms were created or changed only by natural disasters. The earth had once been covered by water (a giant flood), and when it dried out, certain forms took shape. They would remain that way until another natural disaster brought about the next change. Hutton disagreed with this belief. He argued that the earth was a giant heat source with a center that was hot enough to be continually causing changes in rock formation. He also claimed that this process of constant heating, forming, breaking down, and reforming required much more time than previously thought. Therefore, the earth must be millions of years old.

Both claims were hugely controversial since they contradicted biblical beliefs about the age of the earth. In 1795, Hutton published *Theory of the Earth with Proof and Illustrations*. In it, he presented many key theories that are still in use today. As a result, Hutton is remembered as the father of modern geology. He made three important contributions to the field: First, the principle of uniformitarianism stated that rocks form now in the same way that they have always done. The earth's processes are ongoing. In other words, if we study how the earth changes today, we can understand how it changed in the past and how it is likely to further transform in the future. This same principle is used in modern plate tectonics to show us how continents and landforms were created millennia ago. Second, he identified metamorphic rock. Moro had identified two types of rock: sedimentary and volcanic (or igneous). Hutton observed a third type. Sometimes a rock would be exposed to extreme heat or pressure, which would cause changes in the rock's composition and appearance.

The result was a metamorphic (changed) rock. Examples include slate and marble. The former is shale that has been exposed to extreme heat and/or pressure. The latter was once limestone. Third, he explained the rock cycle. As explained above, Hutton believed that rocks took shape but were then broken down or changed to re-form as new types of rock. This cycle is continuous. Forces that can affect and transform the rock include heat, pressure, and natural meteorological factors, such as erosion by wind or rain.

James Hutton died in 1797, but he left behind a body of work that was widely debated for at least another hundred years. His findings influenced many geologists who came after him as well as some in other fields. His observations

In 1872, the HMS *Challenger* embarked on a four-year voyage to explore and chart the bottom of the world's oceans.

about animal husbandry may well have influenced Charles Darwin's theories of natural selection.

Understanding the composition of rocks and how they might have formed was just one part of the giant plate tectonics puzzle that would need to be solved. It was, nevertheless, an important part. The development of geology would eventually combine with physics and several other disciplines to form a more rounded picture.

What about some other parts of the puzzle?

EXPLORING the OCEANS

In 1872, an expedition set off to explore a part of the landscape that had not yet been recorded—the landscape at the bottom of the oceans. The HMS *Challenger* voyage took four years and was the world's first to explore the oceans. Scientists and naval officers visited more than 350 locations, collecting data along the way. So much information was collected that compiling it all took twenty-three years. Ocean depths were measured, along with temperatures, water chemistry, and ocean floor sediment. Since tools to measure ocean depth had not yet been invented, measurements were made via lengths of rope and weights.

During the course of the voyage, researchers discovered that at the bottom of the ocean was a landscape as varied as that above sea level, with mountains, flat plains, and deep trenches. They discovered a mountain range under the Atlantic Ocean and one of the deepest spots of the sea bed in the Pacific. The expedition's findings were a huge step forward for geologists. After all, mountains and trenches were forming beneath the oceans as well as above. Whatever movement was occurring was not limited to what we considered land (that is, above sea level).

FINDING WAYS to MEASURE EARTHQUAKES

What about earthquakes? The ancient Greeks had explained them as pockets of hot air bursting through the earth's surface. In the late nineteenth and early twentieth centuries, a new generation of scholars would come up with better ways to measure and study the tremors.

A seismograph is a device that measures the vibrations of the earth. Most of us today recognize it as a rotating drum device with a pen marking zigzags as the ground shakes. The earliest devices were created in ancient China and consisted of a jar filled with balls. When a tremor occurred, some of the balls dropped into a dish below. Later Italian devices, designed during the sixteenth century, used water or mercury in tubes.

The first modern seismograph was invented by John Milne, an English geologist. In 1875, he traveled to Japan to teach mining. While there, he started to study the nation's volcanoes and earthquakes. He noticed that earthquakes and volcanic activity were not always linked. In fact, there were mountainous parts of Japan with many volcanoes; yet, they seemed to rarely experience earthquakes. At the same time, many of Japan's earthquakes were not centered on or based around a nearby volcano. This led him to rethink certain accepted ideas, but he realized that if more study was to be done, better instruments were needed to measure and record the energy waves of an earthquake. He developed a

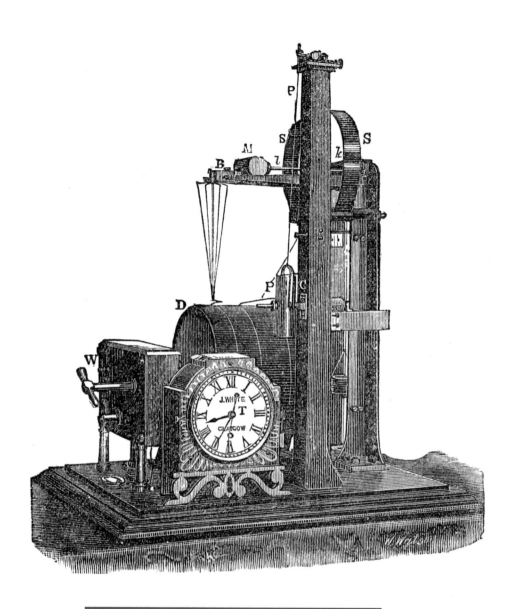

John Milne invented his seismograph after a trip to Japan in 1875 as a means to better chart earthquakes.

horizontal seismograph that used a pendulum to measure movement. Milne placed the devices in seismography stations around the world and compared their findings. Through this, he discovered that the energy waves did not move in the way that everyone thought. Some were fast, and some were slow. Some moved side to side, and others moved back and forth. Geologists would use these findings to determine the different layers of the earth.

A seismograph can detect and record the energy waves of a tremor. But how can we define the size of an earthquake? For this, we need some sort of a scale. One of the earliest to be developed was the Rossi-Forel intensity scale. This scale had ten varying levels, from the level 1 microseismic tremor (something detected only by one seismograph but not by any others) to the level 10 extremely high-intensity tremor, which was defined as "great disaster, ruins, disturbance of the strata, fissures in the ground, rock falls from mountains." This was the first scale to be widely adopted, and it was used from its introduction in 1883 until 1902, when it was replaced by the Mercalli scale.

Like its predecessor, the Mercalli intensity scale relies on qualitative data, such as observations and evaluations of damage. This is not always reliable. In a remote area, the damage may seem less because there are fewer buildings. Yet, the earthquake may have been just as strong as one that destroys a town. Nevertheless, his scale became the standard for measurements. It ranged from scale 1 ("not felt") to scale 12 ("extreme"). Over the years, some alterations were made to the scale. It remained the best available option until Charles Richter and Beno Gutenberg developed their scale in 1935.

While the Mercalli scale measures intensity of an earthquake, the Richter scale measures magnitude. But, what is the difference? The former is descriptive. It is based on what the person grading the earthquake sees—damage to buildings, cracks in the ground, and so on. The Richter scale is a more quantitative measurement. It measures the amount of seismic energy at the source of the earthquake. A Mercalli scale measurement will vary according to how close you are to the epicenter. The closer you are, the stronger it will seem. A Richter scale measurement is focused purely on the epicenter.

Seismologists Richter and Gutenberg developed the scale while working at the California Institute of Technology (known as Caltech) in Pasadena, during the 1930s. The Mercalli scale was too subjective, and so they were looking for something that was more scientifically accurate. They developed a mathematical logarithm that takes into account maximum wave amplitude at a fixed distance from an earthquake's epicenter. The scales range from less than 1.0 (a micro earthquake that is not felt) to 10.0 (an epic earthquake, the likes of which has never been recorded). The Richter scale measures earthquakes in whole numbers and fractions. Each whole number is ten times stronger than the one before it. So, a 4.0 earthquake is ten times the strength of a 3.0. Earthquakes that have a magnitude of 4.5 or higher can generally be detected by sensitive seismographs in other parts of the world. While there are thousands of earthquakes each year, on average only one of a magnitude of 8.0 or higher occurs each year. The highest recorded earthquake using the Richter scale took place in Chile in 1960. It measured 9.5 and caused the deaths of 1,900 people.

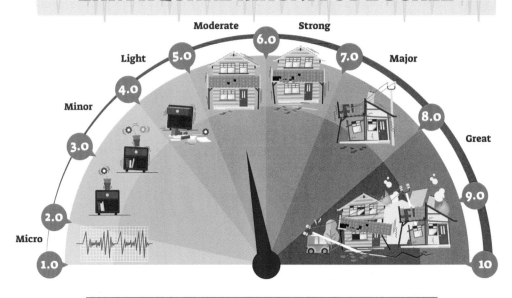

This diagram explains Richter scale seismic activity, from moving furniture to crashing buildings.

Since 2002, the Richter scale has been largely replaced by the moment magnitude scale, which was developed in the 1970s. It uses a similar table of magnitudes as that used in the Richter scale, but the formula used to calculate the scale is different. It also takes advantage of newer technology, since modern seismometers are much more sensitive than those of the 1930s. Scientists can now measure a much broader range of seismic waves and produce a more accurate measurement.

CONTINENTAL DRIFT

Seismology. Oceanography. Paleontology. Physics. Geology. In the field of plate tectonics, all of these separate but related disciplines play a part. It makes sense that the person to make the first significant link in connecting the parts of the puzzle should be someone with a broad range of interdisciplinary interests. Alfred Wegener was an astronomer, meteorologist, and physicist who also had a love of studying maps. It was while looking at maps that he noticed the similarity in the coastlines of some continents. South America and Africa looked as if they could almost fit together. In addition, other studies had noted that fossils of the same creatures from the same time periods had been found on different continents. How could plant seeds and tiny animals have made their way across the vast oceans that separated the continents in such a short time? There were even fossils of ancient tropical plants found in frozen areas so cold that they could not possibly grow there.

Abraham Ortelius

In this chapter, we have learned how Alfred Wegener was partially inspired in his theories by reading about a mapmaker from the sixteenth century. That mapmaker was Abraham Ortelius. Ortelius was born in the Belgian city of Antwerp in 1527. After training as an engraver, he set up his business as a trader in books and maps. Later in life, he was appointed the royal geographer to King Philip II of Spain. He traveled through western Europe, studying and drawing detailed maps from his journeys.

It was his maps that made him famous. In 1564, Ortelius published his first map of the known world. Six years later, he published *Theatrum Orbis Terrarum*, an atlas that contained seventy maps. He continued to publish new atlases for the rest of his life. As he drew his maps, he noticed that the coastlines of America, Europe, and Africa looked as though they might fit together, like pieces of a jigsaw puzzle. He suggested that perhaps the continents were once all joined together and had been torn apart by earthquakes. Several centuries later, Wegener read Ortelius's work and echoed the ideas of a giant landmass separated by massive ruptures. He called this continental drift. Ortelius's ideas would be proven correct but only after scientists had greatly expanded their knowledge of geology.

The commonly accepted theory was that ancient land bridges had once joined the continents. For example, it was believed that there was one land bridge connecting Siberia to Alaska. Another linked Brazil to west Africa. These land bridges had apparently crumbled into the oceans long ago, leaving behind no trace except for the fossil evidence of wildlife. Wegener came up with an alternative theory that would not only explain the formation of the continents but also of mountain ranges.

Wegener believed that instead of being joined by land bridges, the continents were all once part of a giant supercontinent. He named this landmass Pangaea from the Greek, meaning "all lands." According to his estimates, Pangaea had existed roughly 250 million years ago. A few other scientists had suggested something similar in the past. But, they reasoned that parts had sunk into the oceans. Wegener suggested that Pangaea had actually split apart. Over thousands of years, segments had drifted away from each other. He called this movement continental drift.

According to Wegener's theory, Pangaea started to drift apart some two hundred million years ago. At this point, it split into two masses—Laurasia and Gondwanaland. These then split and drifted farther apart to create the seven continents as we now recognize them. Furthermore, as the continents moved through the ocean beds, they encountered some resistance. It was the force of this resistance that caused parts of the continents to compress and rise up, folding in on themselves. This explained how mountain ranges formed. For example, the Himalayas were created as India drifted up into the rest of Asia.

Wegener cited five types of observational evidence to support his theory. These included the following:

- Fossils. As mentioned, he noticed that fossils of the same species were present on different continents. This allowed for several options. Either the exact same species had evolved separately in two different locations or mating pairs had somehow migrated across large stretches of ocean. The more likely scenario was that the lands had once been joined, taking species as they split.

- Interlocking continents. Wegener was not the first to notice how the continents seemed to fit together. Abraham Ortelius made mention of it centuries earlier, as had many others. Oceanographers have since found that these similarities in coastline fit even more accurately at ocean depths, where forces such as erosion have not changed the coastline.

- Geology. Studies of the rock types and formations in west Africa and South America find that they were once part of the same rocky outcrops, formed hundreds of millions of years ago.

- Glaciers. Wegener spent a great deal of time on research expeditions in Greenland, and so the movement of glaciers was something that he was very familiar with. Studying glacial records of continents where they were once probably connected to Antarctica suggests evidence of movement. The alternative explanation would be that half of the globe was covered in an enormous ice sheet. This is known not to be the case given the development of other rock types during the same period.

- Mountains. Wegener found that when he pieced together various landmasses, they formed one solid mountain ridge that stretched from Greenland down through modern-day Scotland, England, Ireland, and Canada.

There seems to have been a great deal of evidence of movement. But, just how did this movement occur? The explanation for this would prove to be the weak point in Wegener's theory. He suggested that the earth's rotation created a centrifugal force. This force was so great that over millennia it pulled everything toward the equator. Pangaea, he claimed, had originally been located near the South Pole. The centrifugal force had caused it to drift apart and move in a northerly direction. He also suggested that gravitational forces from the sun and moon may have caused continental masses to shift to the east or west. These explanations were instantly rejected as implausible. In fact, his entire theory was openly ridiculed. One British geologist said, "It was more than any man who valued his reputation for scientific sanity ought to venture to advocate anything like this theory that Wegener has been able to put forward." Rollin Chamberlin, a geologist at the University of Chicago said that to believe Wegener's hypothesis would mean forgetting "everything we have learned in the last 70 years."

Interestingly, when Wegener was first developing his hypothesis of continental drift, he mentioned the possibility of seafloor spreading, the idea that the ocean floor was slowly being torn apart with new magma rising up into the crevices and cooling. He later abandoned this theory, and it did not appear in his 1915 book about continental drift. As we shall discover in chapter 4, it was this very idea of seafloor spreading that would later be proven by Harry Hess and that would contribute to Wegener's ideas being more widely explored and accepted.

PANGAEA

Pangaea was a supercontinent that existed during the late Paleozoic and early Mesozoic eras.

German scientist Alfred Wegener came up with the theory of continental drift, but he died before he could make further progress in plate tectonics.

CHAPTER 3

The Major Players in Plate Tectonics

The theory of plate tectonics evolved from studies in a range of different scientific disciplines, including geology, mathematics, physics, oceanography, and paleontology. This chapter will provide information about the lives of some scientists who played a major role in the theories of continental drift and plate tectonics.

ALFRED WEGENER

Although Alfred Wegener is widely remembered now as the father of continental drift, he would not live long enough to see wide acceptance of his theories. Yet those theories have contributed so much to our understanding of plate tectonics. During his lifetime, he was much more widely known for his polar expeditions to Greenland.

Alfred Lothar Wegener was born in Berlin, Germany, in 1880. The youngest of five children, he was a good student. After graduating from school as the best in his class, he traveled to Berlin, Heidelberg, and Innsbruck to continue his studies in physics, meteorology, and astronomy. In 1905, he earned his PhD in astronomy from the University of Berlin. However, his interest in meteorology had increased, and he believed his future was in meteorology rather than in the astronomy field. He took a job at a weather station, working alongside his brother. Together, they used weather balloons to study air movement and wind patterns.

The following year, Wegener undertook the first of four expeditions to Greenland. He was appointed the official meteorologist on the two-year research expedition. While there, he built Greenland's first weather station. On his return, he took a position lecturing in physics and meteorology at the University of Marburg, in Germany. In 1911, he wrote his first paper explaining his theory of continental drift, and in January 1912, he gave his first lectures about the idea. Like Abraham Ortelius, several centuries earlier, Wegener noticed that the outlines of the continents seemed to all fit together like a jigsaw puzzle. These smaller pieces could be joined to form one larger landmass. At the time, prevailing belief was that a land bridge had once connected Europe to the Americas. This land bridge had sunk many thousands of years ago. After analyzing fossils and plants found on both sides of the Atlantic Ocean, Wegener noticed many similarities. In fact, many were identical. He suggested that the two continents were once joined and that they had drifted apart. He believed that the continents were all on plates that were constantly moving.

The University of Cambridge in England has been the site of many scientific discoveries. It is also where Dan McKenzie did much of his research on plate tectonics.

Wegener presented his theory of continental drift at the 1926 American Association of Petroleum Geologists in New York. But it was considered controversial, and he was widely ridiculed. Three years later, he wrote what would be his final edition of *The Origin of Continents and Oceans*. Once more, the work promoted the idea of continental drift.

Sadly, Wegener was unable to make any further progress in his studies. In 1930, he embarked on a fourth research trip to Greenland. After being stranded in extreme weather and having run out of supplies, many of the research team died, including Alfred Wegener. During his lifetime, he had suggested that the continental plates moved, but he had been unable to explain why. Still, his work laid the foundations for other researchers a few decades later. The Wegener Medal is now awarded in his memory. Also named after him are the Alfred Wegener Institute for Polar and Marine Research in Germany, an asteroid, a peninsula in Greenland, and craters on both the moon and Mars.

DAN MCKENZIE

Dan Peter McKenzie, the son of a surgeon, was born in Cheltenham, England, in 1942. He was educated at a prestigious private school in London before going on to earn a degree in natural sciences and theoretical physics at the University of Cambridge, in England. By the time he graduated, McKenzie had grown bored with the subject and so he did not do as well as expected in his final exams. However, his tutors saw that he had strong abilities and recommended that he work on a PhD in geophysics. He finished his doctorate in 1966, and he spent the next few years as a visiting scholar at several universities in the United States.

During his graduate studies, McKenzie developed a fascination with convection and the movement of the earth's crust. In 1967, while in the United States, he published the article that would earn him recognition in American geophysical circles. He had been investigating oceanic plates, and he wrote an article with colleague Bob Parker that put forward his theory of plate tectonics and the mechanics behind it. He argued that there were, in fact, two layers in the earth's mantle and that they were both in motion. When the paper was first released, there was some controversy. Just one year earlier, American scholar Jason Morgan had presented on the same topic at a conference. The mathematics both used to reach their conclusions was the same, leading some to suggest that McKenzie had copied Morgan. However, there was no indication that McKenzie was familiar with Morgan's work.

Following the publication of his plate tectonics theory, McKenzie became quite well known in the United States. He received a number of job offers but rejected them all so that he could return to Cambridge. He still teaches there as the Royal Society Professor of Earth Sciences. His research has included seismic activity in Iran and Iceland and the possibility of plate tectonics on other planets. He has received numerous awards and honors over the years, including being appointed a companion of honour by Queen Elizabeth II in 2003.

In 2003, Jason Morgan received the National Medal of Science from President George W. Bush in recognition of his contributions to plate tectonics.

JASON MORGAN

William Jason Morgan was born in Savannah, Georgia, in 1935. After high school, he enrolled in the mechanical engineering program at the Georgia Institute of Technology, in Atlanta. His tutors sensed that he might be better suited to physics than engineering, so they encouraged him to change his major. He did and went on to graduate in 1957 with a BSc in physics. Two years of military service in the United States Navy followed before he entered Princeton University, in New Jersey. He completed his PhD in physics there in 1964. Three years later, he took a job in the Geology Department at Princeton. This is where he would spend his entire career.

Morgan was interested in the fractures on ocean beds and the magnetic anomalies he observed in the areas around them. Combining his observations with his knowledge from the navy and his mathematical skills, Morgan calculated the locations of the various plates. He concluded that the earth's surface is covered by twelve plates. All of these plates are constantly moving. In 1968, he published his plate tectonics theory, based on his research and an earlier presentation he had given. Professor F. A. Dahlen, chair of the Department of Geophysics at Princeton, would later describe Morgan's theory of plate tectonics as "one of the major milestones of US science in the twentieth century." His work and that of Dan McKenzie would provide the foundation for most, if not all, of the geological studies and advancements that have been made since then. Since both Morgan and McKenzie's theories were published almost at the same time, they are both referred to as the codiscoverers of plate tectonics.

Jason Morgan and Critical Mass

> I seriously doubt that I would have worked on these problems if I had not been at Princeton. I'm a strong believer there needs to be a "critical mass": several colleagues all interested in the same problems who frequently converse and are right there to help each other when unanswered questions arise.

This quote from a 2017 interview with Jason Morgan highlights one of the key discoveries to so many breakthroughs in plate tectonics and in many scientific disciplines.

Throughout this chapter, two institutions have been mentioned repeatedly: Princeton University and the Caltech Seismological Laboratory. Both were centers for geological and seismological research.

Morgan understands the importance of having other researchers in related disciplines close by. Their new ideas feed into each other, so conversations can lead to important discoveries. If one person has an idea, there is someone else in the adjacent lab or office who can help to brainstorm and provide a different perspective.

Morgan's comment reinforces the fact that even though a new theory may be named after one particular person, quite often, he or she has been able to make those advancements thanks to collaboration with other like minds.

Very few scientific advancements have ever been made in isolation. Various theories contributed to the recognition of plate tectonics. From Alfred Wegener's theory of continental drift to the understanding of the earth's magnetic field and sea-floor spreading, the theory of plate tectonics has evolved throughout history. Since the development of the theory of plate tectonics, geologists continue to reexamine other aspects of geology. Plate tectonics today is continuously advancing with new technology, enabling scientists to better predict geologic events. The "critical mass" of inquisitive minds was essential to the development of plate tectonics.

In 2003, Morgan retired from a lengthy teaching career at Princeton and spent some time as a visiting scholar at Harvard University, in Cambridge, Massachusetts. He has received multiple awards, including the Alfred Wegener Medal of the European Geosciences Union (1983) and the National Medal of Science of the United States.

Although Wegener, McKenzie, and Morgan are the three principal figures in the development of plate tectonics, there are many others whose work contributed to our understanding of seismic activity. Without these breakthroughs, McKenzie and Morgan would not have been able to reach their conclusions. Some of these researchers are discussed below.

CHARLES RICHTER

Everyone has heard of the Richter scale for measuring earthquakes. But how much do you know about the man who created it? Charles Francis Richter was born in 1900 into a German family in a farming community near Cincinnati, Ohio. After his parents divorced, he and his mother went to live with his grandfather. In 1909, they all moved to Los Angeles, California. Richter's grandfather was very particular about his grandson's education and made sure that he went to some of the best local schools. Young Charles graduated from Stanford University, in California, with a degree in physics in 1920. Eight years later he earned his PhD in theoretical physics at the California Institute of Technology (Caltech), in Pasadena.

During his time in graduate school, Richter developed an interest in earthquakes. After finishing his studies,

Charles Richter, shown here, and Beno Gutenberg worked together to develop a scale for measuring earthquakes. Later in life, Richter worked to help design safer earthquake-proof buildings.

he took a job as a research assistant at the Seismological Laboratory of the Carnegie Institution in Pasadena, California. The lab is now a part of Caltech. Together with fellow seismologist Beno Gutenberg, Richter worked on creating a way to measure and categorize earthquakes. There was a measurement scale already in use, but the pair believed it was not an accurate scientific scale. It was qualitative rather than quantitative. They designed a seismograph that would measure the intensity, or magnitude, of tremors. From this, Richter and Gutenberg came up with the scale that now bears Richter's name.

In 1936, Richter took a teaching position at Caltech. He remained there, teaching physics and seismology, until he retired in 1970. As part of his continuing research, he carefully mapped earthquake zones within the United States. He also spent some time in Japan as a Fulbright scholar. When he returned, he used what he had learned there to help develop better building methods and construction codes in at-risk areas. The city of San Fernando, in California, would later say that thanks to Richter's advice on building safety, the damage caused in the 1971 earthquake that impacted the area was much less than it could have been. The city claims his input saved many lives. Charles Richter died in 1985. The Seismological Society of America now gives an award in his name. The Charles F. Richter Early Career Award honors outstanding contributions to the field by someone who is just starting his or her career.

BENO GUTENBERG

Richter's partner in developing the scale that would carry his name was Beno Gutenberg. Gutenberg was born in 1889, in Darmstadt, Germany. In 1911, he received his PhD in geophysics from the University of Gottingen, in Germany, and he worked at the Geophysical Institute. In 1913, he provided the first accurate calculations for the radius of the earth's core. He was drafted into the German army during World War I and eventually was assigned to the weather forecasting service. After the war, he returned to Darmstadt. His brother had been killed in the war, and his father was struggling to maintain the family soap factory. Gutenberg assisted with the family business while continuing his research. He published a number of papers during this time and was offered an assistant professorship.

Gutenberg was Jewish. As the 1920s progressed and anti-Jewish sentiments grew stronger in Germany, he noticed that his chances for work were limited, despite his academic excellence. So, in 1930, he decided to leave his home country. He moved to Pasadena and worked in the Seismological Laboratory. It was here that he met Charles Richter, and the pair collaborated on developing the earthquake magnitude measuring system. Later in life, Richter would say that the proper name was the Gutenberg-Richter scale. However, Gutenberg disliked dealing with journalists, so his involvement was frequently ignored.

A few years later, he became a professor of geophysics at Caltech. From 1947 to 1958, he served as the director of

Harry Hess used sonar to map the ocean beds and found evidence of seafloor spreading, which would help to prove Wegener's theory of continental drift.

the Seismological Laboratory. He retired in 1958, but he continued to study independently. Gutenberg died in 1960 from pneumonia. Beno Gutenberg received numerous awards and honors, including the Bowie Medal of the American Geophysical Union in 1953; the Prix Charles Lagrange from the Academic Royale des Sciences in 1952; and an honorary degree from the University of Uppsala in 1955.

HARRY HESS

Harry Hammond Hess was born in New York City in 1906. After receiving his PhD from Princeton University he taught at Rutgers University, in New Brunswick, New Jersey, and worked as a research assistant at the Geophysical Institute in Washington, DC, before returning to Princeton to join the faculty. He spent his entire career there, except for one year when he traveled to South Africa as a visiting professor and another year as a visiting professor at the University of Cambridge. From 1950 to 1966, Hess was chair of Princeton's Geology Department.

During World War II, Hess served in the navy as captain of a transport ship. As it happened, the ship was equipped with new sonar technology. Experience with this technology would prove useful in Hess's later research. In fact, he continuously tracked and recorded the ocean floor as he captained his naval vessel around the Pacific Ocean. After the war, he continued his research and teaching. His ongoing surveys of the ocean floor enabled him to develop the concept of seafloor spreading. In 1962, he published *History of Ocean Basins*, which discussed his ideas in greater detail. These findings confirmed Wegener's continental drift theory, so they were an important step forward in the move to understanding plate tectonics.

Also in 1962, President John F. Kennedy asked Hess to chair the Space Science Board. As chairman of the board, he was involved in planning the Apollo missions to the moon, and he was able to analyze some of the rocks brought back from the Apollo 11 mission.

Hess died in 1969, in Massachusetts. He was posthumously awarded NASA's Distinguished Public Service Award, and in 1984, the American Geophysical Union established the Harry H. Hess Medal in his honor. The medal is awarded to "honor outstanding achievements in research of the constitution and evolution of Earth and sister planets."

KIYOO WADATI

The United States has been the center for seismological research. However, some important discoveries were made on the opposite side of the Pacific Ocean, in earthquake-prone Japan. They would later influence the research of scientists in America.

Kiyoo Wadati was born in the Japanese city of Nagoya, in 1902. He graduated from the Imperial University of Tokyo's Institute of Physics in 1925, and then he went to work for the Central Meteorological Observatory of Japan as a seismologist.

While working at the observatory, Wadati developed a theory that the many earthquakes in Japan were the result of movements of plates within the earth's crust. His early work involved studying two earthquakes that had epicenters near each other. He found that although the epicenters were close together on the surface, they originated from different depths within the earth. One occurred about 98,425 feet (30,000 m) from the surface while the other occurred at a depth of more than 984,252 feet (300,000 m).

This discovery led Wadati to study more than a dozen earthquakes in the Honshu region of Japan that occurred between 1924 and 1927. Wadati plotted the data from these earthquakes and was able to define an intermediate and deep earthquake zone off the coast of Japan. He would go on to write up his findings from the study into a paper he released in 1928 about shallow and deep earthquakes. The area of intermediate and deep earthquakes would become known as the Wadati-Benioff zone after another scientist, Hugo Benioff, proved that such a zone exists in each area of the Circum-Pacific region, also known as the Ring of Fire.

Wadati's work would also go on to be a foundation for the work of Charles Richter in developing his scale of earthquake magnitude in 1935. Wadati was placed on the retired list for the Central Meteorological Society from 1929 to 1931 after he came down with a severe case of tuberculosis. However, he did recover and went on to become the director general of the Japan Meteorological Society, a post he would hold from 1956 until his retirement in 1967. Retiring from the Meteorological Society was not the end of his career though. Wadati spent six years as the president of Saitama University and later became president of the Japan Academy. He died in 1995. Among the accolades he received during his lifetime were an imperial gift and prize from the Japan Academy in 1932, in recognition of his discovery of deep earthquakes, and the Order of

Cultural Merits in 1985. He was an honorary member of both the Seismological Society of America and the Royal Astronomical Society.

HUGO BENIOFF

Hugo Benioff's skill at designing earthquake instruments would prove invaluable to his colleagues at the Seismological Laboratory in California. Victor Hugo Benioff was born in 1899 in Los Angeles, California, to immigrant parents. He graduated from Pomona College, in Claremont, California, in 1921, and he wanted to be an astronomer. There was just one problem. He didn't like having to work at night, which made watching stars difficult.

So, he switched his interests to seismology and took an assistant physicist position at the Carnegie Institution Seismological Laboratory. In 1935, he would earn his PhD from Caltech. His mechanical skills allowed him to design several types of the seismometer and the Benioff strain instrument, capable of recording stretching along the earth's surface. As well as designing new instruments, Benioff studied deep earthquake zones in the Pacific Ocean. He advanced Wadati's theories by linking them to subduction. As a result, the zone now bears the names of both scientists. Along with Richter and Gutenberg, Benioff is credited with making the Seismological Laboratory the world's premier earthquake research center.

Ever keen to be inventing, Benioff developed a side hobby of building electronic musical instruments. After all, earthquake vibrations were not much different from sound vibrations. Among those instruments he designed were a

Arthur Holmes used his knowledge of radioactivity to develop a detailed geological time scale of the earth, showing that our planet is much older than was previously believed.

working piano, cello, and violin. In the latter decades of his life, he worked with several famous musicians and a piano company. Hugo Benioff died in 1968.

ARTHUR HOLMES

Geologist Arthur Holmes proved willing to ignore the common consensus that Wegener's continental drift theory was ridiculous, and instead he set out to examine the mechanics that might make such a theory possible. Holmes was born in 1890, in Gateshead, England. He enrolled at London's Royal College of Science to study physics. However, in his second year, he decided to ignore his tutors and take a class in geology. The fascination was instant and from then on, he switched his topic of concentration. After graduating, he traveled to Mozambique, in southeastern Africa, to prospect for minerals in the hope of making his fortune. As it happens, he did not find anything of value. He did contract malaria, and his death was reported to his family in Britain.

Making a full recovery, he returned to England, much to the surprise of his bereaved family. He returned to school, earning his doctorate in 1917. After a failed four-year period in Burma, in southeast Asia, as a geologist for an oil company, he returned home penniless. He took a position as a reader at Durham University, in England, where he would stay until 1943 when he took over as chair of the Geology Department at the University of Edinburgh University, in Scotland.

Holmes had studied radioactivity during his time at the Royal College of Science. He used this knowledge to develop a method of geological dating by measuring radioactive decay. He also created the first geological time scale, which calculated that the earth was far older than was believed. Lastly, Holmes was a strong champion of Alfred Wegener's theory of continental drift. While others ridiculed the theory, Holmes worked on the concept of convection cells in the mantle. This would act as a means of movement for the landmasses above it. This proved to be an important step toward plate tectonics theory.

Holmes retired in 1956 and died in 1965. During his life, he received multiple awards, including the Murchison Medal from the Geological Society of London and the Penrose Medal from the Geological Society of America. There is a crater on Mars named for him.

It is often thought that scientific breakthroughs are made by one or two people working alone in their laboratory. As this chapter shows new developments rely on many different people, often from many different fields of study. The study of the earth requires geologists who understand the rocks, physicists who understand the mechanics of movement, and mathematicians to perform complex calculations. In addition, we see contributions from oceanographers, cartographers, paleontologists, and inventors. Working alone, they create theories. But it is only through coming together that they can achieve what Jason Morgan described as "critical mass." In the case of plate tectonics, that critical mass has forever changed how we understand the earth.

PANGAEA

The supercontinent of Pangaea began to break apart some two hundred million years ago. In doing so, our modern continents were formed.

CHAPTER 4
The Discovery of Plate Tectonics

In 1912, German-born scientist Alfred Wegener began to construct his theory of continental drift. He suggested that the continents had once been joined into a single land mass, Pangaea. Over time they had separated and moved apart. Although he was correct that the continents had moved, he was unable to satisfactorily explain why it had happened. Wegener suggested that the continents moved through the ocean beds, but they are solid rock. How could one mass of rock move through another? There were still gaps in the puzzle. So how did science move from continental drift to plate tectonics?

The theory of continental drift was widely discussed. But without any evidence as to how or why continents might have moved, it was eventually dismissed by most as highly unlikely. Wegener's theories did have at least one strong supporter, however. A British geologist named Arthur Holmes was keen to find the mechanisms that could explain continental drift. He pushed ahead in his research even at a time when doing so might make him subject to ridicule in scientific circles. Holmes suggested that the earth's mantle contained convection cells that helped to move the crust on top of it. The radioactivity in these cells produced heat, which caused rocks within the mantle to rise and sink as they heated and cooled. This would help to explain why and how the plates moved, but Holmes knew that more evidence was needed.

By the 1950s, advances in technology meant that new evidence could be gathered. Much of this technology was developed or improved upon during World War II. Once the war was over, it could be put to other uses. As it happened, many of the answers could be found beneath the sea.

NEW TECHNOLOGY AIDS ADVANCEMENT

One of the first useful developments predates World War II. In 1913, Dr. Reginald Fessenden, a Canadian inventor, was working for a submarine company in Boston. He was attempting to find a method of communicating and detecting objects beneath the sea. A couple of patents for an echolocation device had been registered after the sinking of the *Titanic*. Fessenden's device had some of the same features as those patents. He designed an oscillator that sent out sounds and then used the echoes to measure distance. Soon the device was attached to submarines. However, its low frequency made detecting small objects difficult. Something more accurate was needed. Over the next few decades, adaptations and improvements were made, still using the concept of echolocation. Similar research was being done in Germany. After World War II, German and American technologies would be combined, and the result was called sonar.

Sonar stands for "sound navigation and ranging." It has a wide variety of uses, including communication, detection, and navigation. Used extensively for military purposes, sonar has also proven exceptionally useful in the field of oceanography. Researchers have been able to use it to map the ocean floor, creating detailed maps of variations in the floor surface. Ocean mapping in the 1940s and 1950s showed even more variation than expected. It also raised more questions. It seemed that the layer of sediment at the bottom of the Atlantic Ocean was much thinner than expected. Why was there so little rock accumulation if the oceans were millions or billions of years old? Other discoveries made during surveys of the ocean floor were evidence of reverse magnctism and of seafloor spreading.

Scientists have found evidence that the earth's polarities have reversed at various times throughout the planet's history. They believe it happens every three hundred thousand to seven hundred thousand years. When this happens, if you were facing what we consider your north, your compass would be indicating south! Magma contains magnetic minerals, and when the magma is liquid, those minerals can rotate to face magnetic north. Once the magma cools and hardens into rock, those mineral deposits are stuck in whatever direction they were facing. But how does this help us understand plate movement?

MAPPING the OCEAN FLOOR

When geologists studied the ocean floor, they were able to map areas where reverse magnetism was indicated. This allowed them to estimate the ages of the rocks and of each segment of the ocean bed. What they found was that the rocks in the center of the oceans are younger than those nearest the continents. This provided proof of seafloor spreading. Magma is pushing up through parts of the ocean floor, pushing it apart. The magma hardens to create a new floor. But if it is pushing the beds apart, why doesn't the earth get bigger? Clearly, if new rock is forming in one location, the old rock must be breaking up or undergoing changes.

Harry Hess, a geologist at Princeton University, came up with the term "seafloor spreading." As he studied the changes in the ocean floor using sonar, he realized that Alfred Wegener had been right. This was proof of Wegner's theory of continental drift. As new sea floor was created, older parts were pushed deeper into the trenches, eventually being recycled back into the depths of the crust and the mantle.

At the same time that Hess was exploring the ocean floors, one of the few women in geology at the time was making a major discovery of her own. Marie Tharp was a geologist and ocean cartographer. She and a colleague produced the first scientific map of the Atlantic floor. As a woman, she was not allowed on the research ships. So while a male partner collected data, she did the analysis and calculations back on dry land. She then drew the maps. But her work was often ignored, and her name was left off many papers and publications. In 1952, while mapping the Atlantic Ocean, she discovered the Mid-Atlantic Ridge.

The Mid-Atlantic Ridge, which divides the North American and Eurasian tectonic plates, extends south from Iceland and is more than 10,000 miles (16,093.4 km) long.

The ridge is a mountain chain nearly 10,000 miles (16,093.4 km) long, running along the divergent boundary of the North American and Eurasian plates. It extends underwater from Iceland to the southernmost tip of Africa.

Tharp's discovery reinforced Wegener's theory of continental drift. It proved that the sea floor was spreading. Since her academic partner did not support the idea of continental drift, he dismissed her findings as "girl talk." However, she was able to convince him of her find, and the Mid-Atlantic Ridge joined the growing amount of evidence for continental drift.

The BREAKTHROUGH

In 1966, British geologist Dan McKenzie attended a conference in New York. While there, he listened to a fellow Cambridge graduate who was about to start teaching geology at Princeton. The talk was about seafloor spreading and unusual magnetic observations in the ocean depths. It was the extra piece that Dan McKenzie needed to complete his puzzle.

McKenzie had been wondering how entire plates could move. He was seeking the "how" to complete an understanding of continental drift. Using his own knowledge and what he heard at the lecture, he developed a model. What if each plate was like a solid concrete paving slab? What if these plates all fit together around a sphere, such as the earth's crust? They were all moving because the layers beneath them were moving, so everything was moving relative to those lower layers. The plates on top moved rather like ice floes on the sea.

McKenzie needed to test his theory. With a series of mathematical equations, he was able to calculate the motion of the plates. When he compared this motion to maps charting earthquakes in the North Pacific, he noticed that the two matched exactly. The earthquakes were caused by movement of the plates. This would also explain how the continents had formed—the plates were always moving. Sometimes two would grind along next to each other. Others would jam into each other. But they were always moving.

Meanwhile, in the Geology Department at Princeton University, someone was reaching the same conclusions. Jason Morgan had been fascinated to learn about giant cracks in the bed of the Pacific Ocean. When he charted the cracks, he noticed that they all seemed to show signs of rotation around the same pole. Using mathematics, he was able to calculate the location of the pole and the rotation. Based on his findings, he knew that the ocean floors were a solid material, just like the continents. He tested his formula on the Atlantic Ocean and achieved similar results. Therefore, Morgan concluded, the oceans and continents all sit upon rigid plates that are constantly colliding with each other.

Both McKenzie and Morgan had reached their conclusions independently of each other. McKenzie published his findings first in 1967, and it caused a stir. As it happened, Morgan had spoken about the same topic one year earlier. However, it soon became clear that neither scholar had heard of the other's research. Both are now credited with discovering plate tectonics.

Once the findings were made public, they were quickly accepted. Within just a few years, plate tectonics was being taught in schools around the world. Geologists and physicists realized that this was the final piece of the complex puzzle that so many had tried to explain.

EXPLAINING PLATE TECTONICS

Wegener was correct in his assertion that there was once a giant landmass or supercontinent. Some three hundred million years ago, the earth looked very different to what it looks like now. A large continent called Pangaea covered much of the southern part of the globe. Everything else was water. Roughly two hundred million years ago, Pangaea started to break apart. First, it split into two continents, which we call Laurasia and Gondwanaland. Later, it split into smaller pieces. Some separated while others were crushed together until the continents of today were created. Before the existence of Pangaea, scientists believed several other supercontinents existed, one more than a billion years ago. They probably broke apart and reformed into a new mass over time. The cycle of change is continuous.

What is the earth made of for this to happen? The earth has three layers: crust, mantle, and core. The core is metallic and comprises an inner core and an outer core. The inner, the very center of the earth, is solid. The outer part is liquid. Scientists believe that this part spins with the earth, creating a magnetic field. The mantle is thick, hot semi-solid rock. Around it is the crust. While we may think of the ground beneath us as quite solid, in geological terms the crust is much like an eggshell, delicate and prone to breaking.

The development of the continents happened due to the movement of tectonic plates. The earth is covered with these rigid plates. The exact number is disputed. Some say that there are twelve, whereas others say there are more than this. Geologists know that there are seven major plates: African, Antarctic, Eurasian, Indo-Australia, North American, Pacific, and South American. There are then a number of minor plates, with areas greater than 1 million square kilometers (386,102.2 sq mi) but smaller than 20 million square kilometers (7,722,043.2 sq mi). Examples of these minor plates include the Philippine, Nazca, Arabian, and Somali plates. Finally, there are microplates, but some geologists dispute whether these should even be considered plates in their own right because they are so small. Usually, they are considered parts of the adjacent major plate. The plates vary in size but also in thickness. In some places, they can be more than 200 kilometers (656,168 ft) thick. Other newly formed areas may only be 15 kilometers (49,212.6 ft) or less in thickness. The greater the thickness, the older that segment of the plate.

The hardened tectonic plates are all constantly moving as they shift over the partially melted material of the upper mantle. The rate of movement varies from one plate to another. In some spots, it can be as little as 2.5 centimeters (0.98 inches) per year; in others as much as 15 centimeters (5.9 in) per year. This may not sound very much at all, but over thousands and millions of years, it can add up to significant amounts of movement.

PLATE MOVEMENT

Since the plates are all continually moving, it makes sense that there are occasions when their edges or boundaries jostle against one another much like bumper cars at

PLATE MOVEMENT

Transform

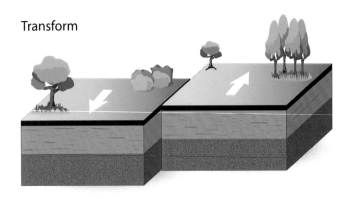

Divergent

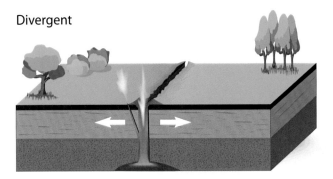

Convergent

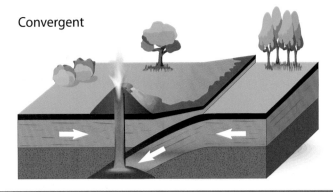

The tectonic plates are constantly in motion, moving closer together, farther apart, or alongside each other.

a fairground. There are many different ways that they can move in relation to the neighboring plate boundary.

When two plates move away from each other, they are called divergent. As the boundaries move apart, new magma flows up from beneath the earth's surface to fill the void and create a new segment of crust. Divergent boundaries along the ocean floor are part of the seafloor spreading process. The plates move apart, and magma rises to fill the gap, creating a ridge as it cools and pushing the plates farther apart. Millions of years ago, seafloor spreading and divergent boundaries separated Saudi Arabia from Africa, creating the Red Sea. Divergence creates rifts in the landscape. Africa's Great Rift Valley is actually a series of separate rifts created thirty-five million years ago by movement along the African, Arabian, and Indian plates. As the rifts grow, it is likely that pieces will break off. The earth's landscape will look very different in another ten million years.

Sometimes two plates collide, causing one to overtake the other. These are called convergent boundaries. The earth is not expanding, so just as new magma burst through in divergent areas, so in other areas plates must shrink to accommodate the new growth. In the case of an oceanic plate colliding with another oceanic plate or with a continental plate, one plate is pushed under the other. This is called subduction. The plate material that is subducted eventually makes its way back down to the mantle where it liquefies into magma. Subduction causes some plates to expand while others shrink, eventually disappearing completely. The Juan de Fuca plate is the smallest

Subduction occurs when one plate is forced beneath another.

tectonic plate. Located between the Pacific and the North American plates, it was once part of a much larger plate. As it is slowly subducted under the North American plate, it is getting smaller and smaller.

When two continental boundaries converge, something else happens. Because both plates are similar in density, they push into each other. This either pushes the land upward into mountain ranges or sideways. Again, remember that this will take place over millions of years. India was once separate from Asia, but the two plates collided, thereby creating the Himalayan mountain range. The Andes, which run along the edge of South America, were also created in this way, with the convergence of the Nazca and the South American plates.

Plates pull apart. Plates push into each other. Sometimes, they move alongside each other. These are called transform boundaries and most commonly occur in the ocean. A few do appear on land, however. The San Andreas Fault in California is an example of a transform boundary caused by the boundaries of the North American and Pacific plates moving alongside each other but in opposite directions.

EARTHQUAKES

Before the development of seismology in the twentieth century, people really didn't understand what caused an earthquake. Now, of course, our understanding has grown.

As the tectonic plates move, they grind against one another. The movement releases large amounts of pressure and energy in the form of waves that course through the ground. These waves are called seismic waves, and the release of them through the ground is what happens during an earthquake. Earthquakes typically take place along fault lines. Faults are cracks in the earth's crust, where one piece of rock will move against another. The major fault lines are located at the boundaries of tectonic plates. Some are barely felt, perhaps a minor shaking beneath our feet. Others are much more severe, opening up cracks, causing buildings to collapse, and giving rise to the risk of a tsunami.

As mentioned in previous chapters, earthquakes are recorded using equipment called seismographs. They are measured according to the moment magnitude scale. The seismic waves radiate from the focus of the earthquake, deep beneath the earth's surface. The epicenter is the point on the earth's surface directly above the focus point. John Milne learned that P (primary) waves move through the earth rather like an accordion, compressing and stretching alternately. Meanwhile, S (secondary) waves shake from side to side. Scientists measure the time both types of waves arrive at the observatory or seismograph and use this to calculate the epicenter of the earthquake. After the P and S waves, other types of waves move across the earth's surface. They move more slowly than those beneath the ground but typically cause more damage.

Thousands of earthquakes happen every year all around the world, but many are so small that they barely register on the moment magnitude scale. Much larger destructive ones happen about once per year. In 2017, one with a magnitude of 8.2 occurred in Mexico City and killed ninety-eight people. In 2015, though not the largest earthquake by magnitude that year, an earthquake in Nepal killed more than eight thousand people.

One of the largest in recent years affected the Tohoku region of Japan. The 9.1 magnitude earthquake in March 2011 killed more than twenty thousand people. It triggered a powerful tsunami and caused massive damage at a nuclear power reactor, which in turn led to explosions and leaks of radioactive material.

The Tohoku earthquake happened in one of the most seismically active parts of the world. The Circum-Pacific Belt, or Ring of Fire as it is more popularly known, extends around the Pacific Ocean, along the western coasts of North and South America, and around the Asian coastline. An estimated 90 percent of the world's earthquakes occur in this region.

In addition to causing a great deal of structural damage to buildings and roads, larger earthquakes can result in dramatic geological changes to a landscape. They can create new cracks along the fault line, widen existing ones, and divert water flows.

TSUNAMI

A tsunami is the result of an earthquake that takes place under water. The epicenter for the 2011 Tohoku earthquake was about 40 miles (64.4 km) off the coast of Japan. The shaking of seismic waves sends ripples through the ocean. These ripples get larger until giant walls of water move toward land with incredible speed and power. When they reach land, they can send a deluge of water for many miles inland, causing massive damage.

VOLCANOES

There are approximately 452 volcanoes along the Circum-Pacific Belt, which adds to the seismic activity in the area. Are volcanoes and earthquakes linked? If so, how?

Volcanoes are openings in the earth's crust that provide a means of escape for gases, lava, and ash from the mantle below. Magma is the name given to molten rock when it is underground. Once the volcano erupts and the magma reaches the surface, it becomes lava. Thick streams of it ooze down the sides of the volcano, eventually hardening. In particularly forceful eruptions, it may shoot into the air along with clouds of ash and smoke.

Like earthquakes, volcanoes are found mostly near tectonic plate boundaries. They can lay dormant for many years, centuries even, before rumbling back to life and entering an active stage. Also, like earthquakes, some volcanic eruptions are minimal, just a small puff of smoke and little more. Others are violent displays of fire that can cause damage for miles around.

If we look at the construction of a volcano, we will see the main vent that leads from the mantle to the earth's surface. Deep within the rock along the vent is a magma chamber. This is where the molten rock from the mantle collects. As the pressure reaches a critical mass, an eruption occurs. The magma is forced upward through the main vent along with gases, ash, and other debris. There may also be side vents in the volcano, where magma has forced its way through cracks in the sedimentary rock layers. Magma can also escape from these side vents, sometimes creating smaller peaks as it hardens. As it cools, the lava hardens into sheets of rock that we often refer to as lava plains, or lava fields.

Sometimes, an eruption will occur with such force that a large chunk of the volcano will explode, leaving behind a very different crater when things cool down.

Even though both are seismic events and occur in the same global regions, volcanic eruptions and earthquakes are not always related. It is possible to have one without the other. On the other hand, there are occasions when the pressure of the magma trying to escape is so great that it triggers an earthquake. Alternatively, an earthquake near a volcano that is showing some signs of activity may set off a larger eruption.

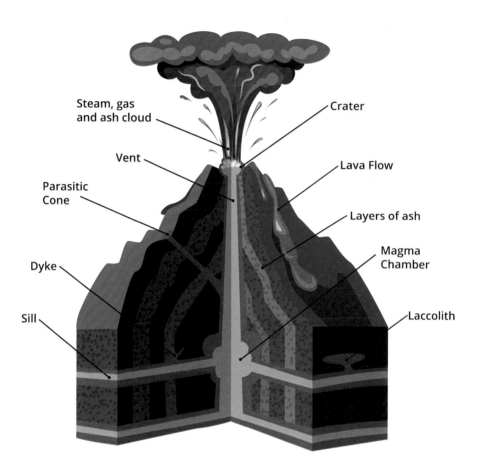

This cross-section diagram shows the various parts of a volcano, including the main vent through which ash, steam, and lava escape.

Hawaii is an active volcanic zone even though it lies in the middle of a plate, something that scientists are still exploring and attempting to understand.

Our modern understanding of earthquakes, volcanoes, and how they occur is due to plate tectonics theory and the work of Dan McKenzie, Jason Morgan, and their colleagues. Yet, it is important to remember that plate tectonics theory is still relatively new—barely fifty years old. There is still more research to be done in some areas.

McKenzie and Morgan have shown us how plate tectonics works. The next question to be answered is how and when it all began. Recent studies suggest that plate tectonics began an estimated 3.2 billion years ago, roughly two-thirds of the way through the earth's existence.

Researchers think that changing temperatures within the earth's mantle may have brought about the beginning of plate tectonics, but they are keen to understand why the temperature changed and how things were set in motion.

Another puzzle is related to so-called hot spots. Plate tectonics allows us to understand how volcanoes can form along plate boundaries. However, there are some volcanic areas, such as Hawaii, that exist in the center of a plate.

In 1963, geophysicist John Tuzo Wilson explained these as mantle plumes. He argued that small, excessively hot spots exist in the mantle below the tectonic plates. The heat becomes so great that the plate above it melts in places, allowing for repeated volcanic eruptions that over time build up on the sea bed, to create the Hawaiian Islands for example. Not all scientists agree with Wilson's theory. Some argue instead that this results from weaknesses within the plate as they spread apart.

Clearly, there is more interesting work to be done in the field of plate tectonics. In the next chapter, we will explore the influence of plate tectonics.

Supercomputers Helping to Model Plate Tectonics

Advances in computer technology in the last decade have allowed researchers to make leaps forward in studying and modeling plate tectonics.

In 2010, a group of computer scientists at the University of Texas at Austin joined forces with geophysicists at Caltech Seismology Laboratory. Together, they were able to create a new supercomputing model that shows a much more detailed view of the earth's plates than ever before.

In the past, modeling plate tectonics on a global scale was almost impossible. The different patterns of flow within the mantle required so many different equations and algorithms that multiple computers were needed to perform the calculations. Even then, there were problems getting all of the computers to work together.

Now, it is possible to use a supercomputer using thousands of processors. The computer scientists have also designed a series of new algorithms that all work in conjunction with each other. The result is a huge step forward for those who study plate movement, earthquakes, and volcanoes.

The new model can simulate mantle flow. This allows researchers to study movement patterns and predict the movement of plates. Because of the level of detail, even movement at individual points along fault lines can be modeled.

Scientists were also surprised to learn new things from the computer model. For example, they have been able to study the energy released during plate movement and have learned that it is not as much as they once thought. Instead, much of the energy is released deep beneath the earth's surface.

Lijun Liu, professor of geology at the University of Illinois, and his team are also using technology to answer questions surrounding plate tectonics. Liu and his team are looking at the processes behind the earth's evolution. The team's data-centric model uses the Blue Waters supercomputer at the National Center for Supercomputing Applications at Illinois. The supercomputer applies complex four-dimensional data-oriented geodynamic models.

The Transamerica Pyramid in San Francisco was designed to be able to survive a strong earthquake.

CHAPTER 5

The Influence of Plate Tectonics

In previous chapters, we looked at the history of the study of plate tectonics. You were introduced to many of the key figures that helped build our understanding of the science behind the movement of the earth's crust and how that movement affects the world around us. Now let's look how at this understanding impacts us today and where it may take us in the future.

EARTHQUAKE-PROOF BUILDINGS?

The movement of the plates in the earth's crust has many consequences for the world around us. One of the most significant of those consequences is the occurrence of earthquakes. As mentioned in previous chapters, thousands of earthquakes take place each year around the globe. These seismic events range from those that can only be detected with special instruments to earthquakes that can topple buildings and cause massive destruction.

In 1906, a 7.9 magnitude earthquake struck San Francisco, killing hundreds and possibly thousands of people. Some buildings that did not collapse were left leaning, such as those shown here.

Despite all the advances in the study of plate tectonics, there is still no way to accurately predict when and where an earthquake will occur. The best that science can do at this time is to forecast where an earthquake is likely to happen and within a certain time frame. This is based on our understanding of the geological forces at work and the history of seismic events in the region. Forecasting the probability of an earthquake can include expectations of the event's magnitude, but there is no way to predict definitely when an earthquake will happen or how powerful it will be.

Though we cannot predict when or where earthquakes will happen, our understanding of plate tectonics and the forces at work within the earth's crust have helped to prepare us in other ways. Architects and engineers have been working for years to develop earthquake-proof buildings. Charles Richter used his knowledge gathered in Japan to assist engineers with construction designs in California. This work is extremely important as many of the world's largest cities are built in or near active seismic zones. But what does it mean for a building to be earthquake-proof?

The first thing to understand about earthquake-proof buildings is that, just as the *Titanic* was called an "unsinkable" ship, no building is truly protected against an earthquake. However, the technology for earthquake-resistant buildings has made significant advances to minimize damage.

Predicting an Earthquake

To predict an earthquake, the United States Geological Survey (USGS) says that three key pieces of information are needed: We need to know exactly when and exactly where the event will happen. We also need to know the magnitude of the event.

It is not possible to predict all three. However, scientists can use observations to make guesses about future activity. For example, volcanologists can say that a group of earthquakes might indicate underground seismic activity. Laser beams can also be used to detect plate movement. A seismometer is used to analyze the vibrations in the earth's crust. Because radon gas escapes from cracks in the earth's crust, levels of the gas can also be monitored.

They might also use the Global Positioning System (also known as GPS) to locate and measure bulges in the ground near a volcano. Both of these could be signs that magma is gathering and pressure may soon force an eruption. But they have no way of knowing exactly when that eruption might occur.

Geologists can also predict the probability of a future earthquake. According to the USGS, the likelihood of an earthquake in the San Francisco area within the next thirty years is 67 percent. But how useful is that information?

Although it is very vague, such data can help areas that are more likely to suffer from earthquakes make some preparations, such as making sure that buildings are safe.

Early warning systems, such as those used in Mexico and Japan, can detect movement as soon as it begins, allowing for surrounding areas to be notified. Even a few seconds or minutes can be enough time to seek shelter.

Hopefully, as technology advances, it will one day be possible to predict earthquakes and volcanic eruptions with more accuracy so that people might have hours or even days to evacuate.

Japan is a country known for its beauty and its unique style of architecture. Before the turn of the twentieth century and the Industrial Revolution in Japan, many of the country's buildings were made of light wood with paper walls and doors. One of the key reasons for this is that Japan sits on an extremely active seismic zone. This means that the country experiences thousands of earthquakes each year. Though most of them are not even noticeable, those like the Kobe earthquake (also known as the great Hanshin earthquake) of 1995 and the Tohoku earthquake and tsunami of 2011 are vivid examples of the power that some of these earthquakes can have. To deal with the threat of earthquakes, Japanese architecture historically relied on lighter materials designed to move with earthquakes. If the buildings did collapse, they could be easily rebuilt.

Modern techniques take the lessons learned from the science of plate tectonics and apply them to how buildings are made. Take for example the Mori Tower in Tokyo. Standing 781 feet (238 m) tall, the designers built it to be "a city to escape into rather than a city from which people run away." To do this, they included many earthquake-resistant design features. The building contains 192 giant fluid-filled shock absorbers. In the event of an earthquake, the thick oil inside these shock absorbers sloshes back and forth to counter the swaying motion of the earth. This helps the building to remain stable even as the ground around it trembles.

Another example of earthquake-resistant design is the Shanghai Tower in Shanghai, China, standing at 2,000 feet (609.6 m). Like Japan, the Shanghai area is seismically active. The ground in Shanghai is also mostly made of hard-packed clay, which is not ideal for a solid building foundation. To counter these problems, the designers started from the ground up. Actually, they started from the ground down. They drove 980 foundation piles into the ground, some of them to a depth of almost 300 feet (91.4 m). Then, they poured more than 2 million cubic feet (56,633.7 cubic meters) of reinforced concrete. This formed a foundation mat 20 feet (6.1 m) thick for the building to stand on. Another key feature of the building's earthquake-resistant design is the inclusion of a 1,000-ton (907.1 metric tons) tuned mass damper. This giant device acts in the same way as the oil shock absorbers of the Mori Tower. The damper moves against the forces of an earthquake, countering the sway generated by the shaking ground. The damper also features its own shock absorbers to prevent it from moving too far or too fast when the building sways.

Both the Mori Tower and the Shanghai Tower are recent examples of earthquake-proof designs. The 853-foot-tall (260 m) Transamerica Pyramid in San Francisco is nearly fifty years old. Despite its age, it has its own features to protect it from seismic events. Knowing that the Bay Area is likely to experience more massive earthquakes like the one that leveled the city in 1906, architects designed the Transamerica Pyramid to withstand a high magnitude earthquake.

The building is built on foundation 55 feet (260 m) deep that is made of concrete and steel. It is designed not to resist the movement of the ground but to move with it in the event of an earthquake. The building's exterior is also made of a specially made quartz aggregate. That in turn is supported by reinforcing rods at four points on each level of the building. This design allowed San Francisco's Transamerica Pyramid to withstand the force of the magnitude 6.9 Loma Prieta earthquake of 1989. The building shook for more than a minute, and the top floors swayed as much as a foot from side to side. However, when the earthquake ended, the building had suffered no structural damage.

PLATE TECTONICS on OTHER PLANETS?

Of course, the study of plate tectonics has led to more than just improvements in building design to resist earthquakes. It has also resulted in advances that go beyond the earth and into space exploration!

The National Aeronautics and Space Administration (NASA) has been using the knowledge we've gained in studying the earth's plate tectonics to help us better understand things happening right here in our own solar system. Study of some of the earth's nearest neighbors in space—our moon, Mars, and Mercury—has revealed that all of them most likely once had active volcanic systems. These systems are now dormant. It is heat, or at least the release of heat, from within the earth's crust that fuels plate tectonic movement. The moon, Mars, and Mercury are all smaller than the earth, and scientists believe that their size allowed the heat within them to escape completely, thereby resulting in the end of any kind of tectonic movement.

Interestingly though, the earth is not the only celestial body in our solar system with active plate tectonics and the resulting seismic activity. Io (pronounced "eye-oh") is one of the planet Jupiter's moons. It is also the only other place

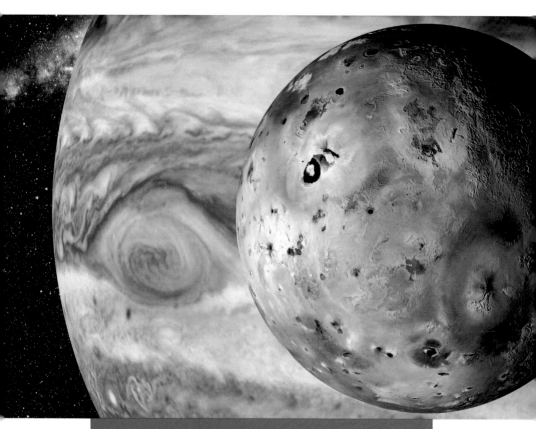

Scientists are also studying plate tectonics in space. Jupiter's moon Io is known to have some very active volcanoes.

we know of in the solar system to have active volcanoes. Scientists had speculated about the existence of volcanoes on Io for years, but it wasn't until the *Voyager* space probe sent back data showing volcanic activity as it passed near the moon in 1979 that scientists had actual proof.

Further study through the Hubble Space Telescope and other space probes has revealed an active volcano system on the moon. While geologists are not yet able to visit Io to make firsthand observations, their understanding of plate tectonics has allowed them to create a theory about what is happening on, and under, the surface of Io. From the data they have received, geologists believe that there is a sea of superheated magma mixed with solid rock under the surface of Io. The moon's orbit is not round, but oval. This elliptical orbit means that at times Io is closer to Jupiter and feels the pull of the massive planet's gravity much more than other times. In the same way that the orbit of the earth's moon affects the ocean tides, the magma ocean below Io's surface is pulled by Jupiter's gravity. Scientists have observed that Io's surface bulges during this process, sometimes more than 328 feet (100 m). This movement generates heat that keeps the magma from cooling and turning to solid rock. In fact, Io's volcanic eruptions mean that the normally cold surface of the planet (averaging -202 degrees Fahrenheit [-130 degrees Celsius]) can have pockets of extreme heat, some as hot as 3,000 degrees Fahrenheit (1,649°C). This is why Io is sometimes called a celestial body of fire and ice.

Though no probes are planned to be sent directly to study Io's surface directly, it is hoped that future missions to study Jupiter and its other moons will allow for closer observation of the moon. Since Io's volcanoes sometimes erupt with enough force to send material 656,168 feet (200 km) or more into space (that's twice the height of the earth's atmosphere), it might be possible for a probe to collect samples of Io's inner material from simply flying near the moon, without ever landing!

Not to be deterred from its study of Io, in 2012, the USGS released a detailed map of Io's surface. The map was a composite creation made using data from the *Voyager 1* and *Voyager 2* space probes as well as from observations through the Hubble Space Telescope. The map was the first to include all of Io's surface. It is also an aid in studying volcanic activity on the moon.

A CONSTANTLY CHANGING EARTH

Geologists have often said that understanding the way the earth has changed in the past can help us to understand the present. It can also help us to forecast what might happen in the future. Plate tectonics means that the earth is constantly changing in new and exciting ways.

As we lose some islands and pieces of coastal land to erosion and rising sea levels, some new islands have appeared as a result of volcanic activity. In 2014, an underwater volcano near Tonga erupted and is now a small island in its own right. Called Hunga Tonga, the island has a growing bird population. Meanwhile, elsewhere in the Ring of Fire, the tiny Japanese island of Nishinoshima, south of Tokyo, is growing. An eruption from a nearby undersea volcano in 2013 created a small island nearby. The two have

The lava lake of Erta Ale is part of East Africa's Great Rift Valley.

now merged, and Nishinoshima is still growing as more lava pours forth. It has doubled in size and shows signs of a flourishing new ecosystem. Two more islands appeared in the Red Sea where the African and Arabian tectonic plates meet. Both have now disappeared back into the sea, but geologists expect to see more activity in this volatile area.

In the 1960s, a rare lava lake was found in northeastern Ethiopia. It is part of a volcano called Erta Ale. Eruptions are common, and since 2005, it has been continuously active. Erta Ale is part of the East African Rift system, which is part of the larger 6,000-mile (9,656.1 km) Great African Rift Valley. Because of its remote location and the political instability of the area, Erta Ale is not an easy one for geologists to investigate. This part of the African Rift lies at the junction of the African and Arabian plates. As the two pull apart, a new plate is forming. Geologists have already named it the Somalian plate.

The area gained worldwide attention in 2005 when a giant crack appeared in the earth's surface following a nearby volcanic eruption. Rather than erupting through the surface, magma was diverted along an underground crack. As it cooled, it hardened into a wedge, which ripped the ground apart. The resulting rift is 35 miles (56.3 km) long and 20 feet (6.1 m) wide in some places. Researchers believe that the crack may eventually reach the Red Sea. At that point it will fill, creating a new sea. Ethiopia will split apart from mainland Africa. Don't expect this to happen soon, though. Such a process is likely to take between four and ten million years. In the meantime, scientists are excited as it is an opportunity to learn more about how ocean ridges might form.

WHAT LIES AHEAD

Jumping ahead several million years, we may find that the shape of east Africa has changed considerably with Ethiopia a separate continent and a new sea. But in what other ways can we expect plate tectonics to change the earth in the future?

Computer models can use what we know about past plate tectonic movement to predict what might happen in the future. If the plates continue their current rates of movement, in another 250 million years, the continents will have all collided into one another to create a new supercontinent, Pangaea Ultima.

Australia will likely continue making its way northward until it collides with and merges into Asia. In a similar fashion, Africa will split into several smaller pieces and will also move northward. As it does so, the Mediterranean Sea will disappear. Northern Africa will merge into southern Europe, creating a mountain ridge along Greece, Italy, and Spain. This is rather like the way the Himalayas formed millions of years ago when India and Asia collided.

As for the Atlantic Ocean, at first it will expand, causing much of the east coasts of North and South America to submerge beneath. Then, it will reverse its course, shrinking as the Americas move to merge into the newly forming supercontinent. Eventually, the mighty Atlantic Ocean may be little more than an inland sea.

Of course, this is all based on current and historical observations. If scientists are able to uncover the mysteries that started plate tectonics more than three billion years ago, they may detect possible changes to come. For example, if the temperature of the mantle or the core

Scientists are continuing to develop computer models of how plate tectonics will change the earth in years to come and to develop buildings that can withstand these changes.

changes, how might that impact the plates? Many questions remain unanswered.

In 1912, when Alfred Wegener developed his theory of continental drift, he knew that he was still lacking some key information that would explain why and how the continents drifted apart. He probably did not expect the open hostility and ridicule that he would receive from much of the scientific community when he presented his theory. Although he hoped to prove his critics wrong, he died before he was able to find the answers he was looking for.

Throughout history, many scientific pioneers have faced ridicule or punishment for their theories. Galileo was charged with heresy and sentenced to home imprisonment for suggesting that the earth revolved around the sun. Even Albert Einstein died believing that several of his theories were wrong, only for them to be proven right later. Wegener never stopped doubting that he would find the explanation for continental drift.

Fortunately, there were some who saw promise in Wegener's ideas. They worked to fill in the missing pieces. The results of many years of research finally came together in the late 1960s as researchers in both the United Kingdom and the United States created a complete picture of plate tectonics.

The findings of Jason Morgan and Dan McKenzie revolutionized the field of geology and are still considered some of the most important scientific advances of the twentieth century. Thanks to these scientists and researchers, we know how a volcano erupts, why the ground shakes, and what created the continents. We also recognize that they are still changing and will continue to do so. Finally, we know that these discoveries are only the beginning of our understanding of plate tectonics.

A knowledge of plate tectonics has allowed for steps to be taken to protect people. While we cannot prevent or avoid seismic events entirely, we now have the knowledge to design safer buildings and to create warning systems that can save lives. Supercomputer models are also enabling geologists to better understand the forces behind plate tectonics.

Geologists continue to explore plate tectonics, delving far into space in search of other examples. Meanwhile, here on the earth, we can watch the changes that are taking place in Africa and ask ourselves, "How will plate tectonics affect the way the planet looks in millions of years to come?" Questions like this will lead us to discover even more about plate tectonics in the years to come.

Chronology

200 million BCE
    The landmass Pangaea splits into two landmasses: Gondwanaland and Laurasia. From there, it continues to split until the current continents are formed.

500 BCE
    Anaxagoras explains his theory of the universe.

350 BCE
    Aristotle presents evidence that the earth is round.

ca. 200 BCE
    Eratosthenes calculates the radius and circumference of the earth.

20 CE   Strabo writes first known geography book and suggests a link between earthquakes and volcanic eruptions.

79 CE   Mount Vesuvius erupts, destroying the towns of Pompeii and Herculaneum.

ca. 1480   Leonardo da Vinci begins collecting and studying fossils.

1564   Abraham Ortelius produces the first atlas of the known world.

1669   Nicolaus Steno publishes his book in which he lays the groundwork for future geological study.

1795   James Hutton publishes his work, cementing his reputation as the father of modern geology.

1872   The HMS *Challenger* embarks on its four-year scientific voyage.

1880   John Milne designs his first seismograph.

1902   The Mercalli intensity scale is widely adopted.

1906   The San Francisco earthquake had an estimated moment magnitude of 7.9.

1912   Alfred Wegener first proposes his theory of continental drift.

1913   Reginald Fessenden invents sonar.

1915   Wegener publishes his first book explaining his theory of continental drift

1928   Kiyoo Wadati publishes his paper explaining shallow and deep earthquakes.

1935   The Richter-Gutenberg scale replaces the Mercalli scale for measuring the intensity of earthquakes.

1952   Marie Tharp discovers the Mid-Atlantic Ocean Ridge.

1962   Harry Hess publishes the findings from his study of ocean floors, providing evidence of seafloor spreading.

1967   Dan McKenzie publishes his article about plate tectonic theory.

1968   Jason Morgan publishes his paper about plate tectonic theory, based upon a talk he gave in 1966.

1979   The space probe *Voyager* sends back images confirming the presence of volcanic activity on Io.

1985   An 8.0 intensity earthquake strikes Mexico City, causing more than five thousand deaths.

2002 The moment magnitude scale replaces the Richter-Gutenberg scale.

2005 A giant crack opens in the earth's surface in Ethiopia.

2010 A new supercomputer enables the most detailed modeling of plate tectonics to date.

2011 A 9.0 intensity earthquake occurs on the coast of Japan, triggering tsunamis. More than fifteen thousand people are killed, and a nuclear power plant is breached, causing extensive damage to the region.

2015 A 7.8 intensity earthquake strikes Nepal, killing nearly nine thousand people and injuring more than twenty-two thousand.

2018 An earthquake measuring 7.2 hits Mexico City.

Glossary

centrifugal force An apparent force that acts outward on a body moving around a center, arising from the body's inertia.

continental drift Theory put forward by Alfred Wegener that proposed that all the continents were once one giant landmass and that they had drifted apart from each other.

convection The transfer of heat via movement of gases or liquids.

convergence The movement of two objects toward each other.

core The innermost part of the earth.

crust The outermost and thinnest layer of the earth.

divergence The movement of two objects away from each other.

earthquake A violent shaking of the ground caused by movements in the earth's plates.

echolocation The use of reflected sound waves to determine the location of an object.

epicenter The point on the earth's surface that is directly above the focal point of an earthquake.

erosion The gradual wearing away of rock caused by natural elements such as wind, rain, and air.

fault A crack or fracture within the earth's crust.

fossil The preserved remains of an organism within rock.

geology The study of the earth, its rocks, and how they change over time.

ichnology The study of fossilized footprints and tracks.

igneous A type of rock formed from cooled lava.

land bridge A strip of land once believed to have connected continents to each other.

lava Molten rock that has erupted from a volcano onto the earth's surface.

law of superposition A basic law of geochronology, stating that in any undisturbed sequence of rocks deposited in layers, the youngest layer is on top and the oldest on bottom, each layer being younger than the one beneath it and older than the one above it.

magma The hot molten material beneath the earth's crust that becomes lava.

mantle The layer of the earth that is between the crust and the core.

metamorphic Rock that has changed its composition as the result of exposure to extreme pressure or heat.

paleontology The scientific study of fossils.

Pangaea The name given to the landmass from which the continents formed. The word comes from the Greek meaning "all lands."

plate tectonics The scientific theory that explains the earth's crust and its movements.

pyroclastic Formed by hot volcanic fragments hurled into the air during an eruption.

rift A geological term for a crack within the crust.

sedimentary Rock that is formed when soil and sediment are compressed into layers.

seismic Related to the earth's crust and earthquake activity.

seismometer An instrument capable of detecting and recording earthquakes. Also called a seismograph.

sonar A method of detecting objects based on echolocation. It uses sound pulses and measures the amount of time for them to bounce back.

subduction The movement of one tectonic plate beneath another at the point of convergence.

theory An attempt to explain something when further research is needed to provide concrete evidence.

transform boundary Places where two tectonic plates slide past each other rather than colliding or pulling apart.

trench A long, narrow ditch typically found on the ocean floor.

tsunami A massive and destructive sea wave caused by an underwater earthquake.

volcano An opening in the ground, usually a mountain, through which lava and gases burst from the earth's mantle.

Further Information

BOOKS

Allaby, Michael. *Earth Science: A Scientific History of the Solid Earth*. New York: Facts on File, 2009.

Anderson, Michael, ed. *Investigating Plate Tectonics, Earthquakes, and Volcanoes*. New York: Britannica Educational Publishing, 2012.

Gosman, Gillian. *What Do You Know About Plate Tectonics?* New York: Rosen Publishing Group, 2014.

Molnar, Peter. *Plate Tectonics: A Very Short Introduction*. Oxford, UK: Oxford University Press, 2015.

Saunders, Craig. *What's the Theory of Plate Tectonics?* New York: Crabtree Publishing, 2011.

WEBSITES

American Geosciences Institute
https://www.americangeosciences.org

The American Geosciences Institute aims to represent and serve the geoscience community by providing information to connect the earth, science, and people. It provides a range of publications and education programs.

McKenzie Archive
https://www.mckenziearchive.org

Organized by the Geological Society of London, this archive contains a complete history of Dan McKenzie's work as well as interviews and a timeline of his work on plate tectonics.

National Geographic Plate Tectonics
https://www.nationalgeographic.com/science/earth/the-dynamic-earth/plate-tectonics

This reference guide provides a brief overview of plate tectonics and the types of plate movement. There are also links to a collection of images and video footage.

United States Geological Survey
https://www.usgs.gov

The website for the USGS provides real-time data about current observations, recent earthquakes, water conditions, and more. It also explains the science behind various geological topics as well as ongoing research and new findings.

Bibliography

Ananthaswamy, Anil. "Dan McKenzie: The Man Who Made Earth Move." *New Scientist*, November 2017. https://www.newscientist.com/article/mg23631530-600-dan-mackenzie-the-man-who-made-earth-move/.

Blakemore, Erin. "Seeing Is Believing: How Marie Tharp Changed Geology Forever." *Smithsonian*, August 30, 2016. https://www.smithsonianmag.com/history/seeing-believing-how-marie-tharp-changed-geology-forever-180960192/.

Bressan, David. "A Short History of Earthquakes in Japan." *Scientific American*, March 11, 2012. https://blogs.scientificamerican.com/history-of-geology/a-short-history-of-earthquakes-in-japan/#.

Brumbaugh, David S. *Earthquakes: Science and Society*. London: Pearson, 2009.

Chester, Roy. *Furnace of Creation: A Journey to the Birthplace of Earthquakes, Volcanoes, and Tsunamis*. New York: AMACOM, 2008.

Fountain, Henry. *The Great Quake: How the Biggest Earthquake in North America Changed Our Understanding of the Planet*. New York: Broadway Books, 2018.

Francis, Peter, and Clive Oppenheimer. *Volcanoes*. Oxford, UK: Oxford University Press, 2003.

Frankel, Henry. "The Development of Plate Tectonics by J. Morgan and Dan McKenzie." *Terra Nova* 2 (1990): 202–214. doi:10.1111/j.1365-3121.1990.tb00067.x.

Frisch, Wolfgang, and Martin Meschede. *Plate Tectonics: Continental Drift and Mountain Building*. New York: Springer, 2011.

Furgang, Kathy. *Everything Volcanoes and Earthquakes*. Washington, DC: National Geographic Society, 2013.

Gallant, Roy A. *Plates: Restless Earth*. New York: Benchmark Books, 2003.

Graham, Ian. *Tsunami: Perspectives on Tsunami Disasters*. Portsmouth, NH: Heinemann, 2014.

Greene, Mott T. *Alfred Wegener: Science, Exploration, and the Theory of Continental Drift*. Baltimore: Johns Hopkins University Press, 2015.

Griggs, Mary Beth. "Volcanic Eruptions Are Incredibly Hard to Predict. Here's Why." *Popular Science*, November 27, 2017. https://www.popsci.com/predict-volcanic-eruption.

Hogenboom, Melissa. "We Have Known that Earth Is Round for Over 2,000 Years." BBC Earth, January 26, 2016. http://www.bbc.com/earth/story/20160126-how-we-know-earth-is-round.

Kearey, Philip, Keith A. Klepeis, and Frederick J. Vine. *Global Tectonics*. Hoboken, NJ: Wiley-Blackwell, 2009.

Knopoff, Leon. *Beno Gutenberg, 1889–1960*. Washington, DC: National Academies Press, 1999. http://www.nasonline.org/publications/biographical-memoirs/memoir-pdfs/gutenberg-beno.pdf.

Livermore, Roy. *The Tectonic Plates Are Moving!* Oxford, UK: Oxford University Press, 2018.

McKenzie, Dan P., and Robert L. Parker. "The North Pacific: An Example of Tectonics on a Sphere." *Nature* 216 (1967): 1276-1280. doi:10.1038/2161276a0.

Molnar, Peter. *Plate Tectonics: A Very Short Introduction*. Oxford, UK: Oxford University Press, 2015.

Morgan, W. Jason. "Rises, Trenches, Great Faults, and Crustal Blocks." *Journal of Geophysical Research* 73 (1968): 1959-1982. http://www.mantleplumes.org/WebDocuments/Morgan1968.pdf.

"New View of Tectonic Plates: Computer Modeling of Earth's Mantle Flow, Plate Motions, and Fault Zones." Science Daily, August 30, 2010. https://www.sciencedaily.com/releases/2010/08/100827092828.htm.

Redfern, Ron. *Origins: The Evolution of Continents, Oceans, and Life.* Norman, OK: University of Oklahoma Press, 2001.

Switek, Brian. "Leonardo da Vinci: Paleontology Pioneer." *Smithsonian*, June 11, 2010. https://www.smithsonianmag.com/science-nature/leonardo-da-vinci-paleontology-pioneer-1-73326275/.

Vuik, Kees, "Is It Finally Possible to Predict Earthquakes?" *Guardian*, May 20, 2015. https://www.theguardian.com/global-development-professionals-network/2015/may/20/is-it-finally-possible-to-predict-earthquakes.

Index

Page numbers in **boldface** are illustrations.

Anaxagoras, 11–13, 16
ancient beliefs and myths, 6, 8, 11–20, 36
Aristotle, 13, 20

Benioff, Hugo, 65–68
buildings, earthquake-proof, 60, 95–100, 109
Bush, George W., **54**

Cambridge, University of, **51**, 52–53, 58, 63, 76
Circum-Pacific Belt, 65, 84–85
construction methods, improved and earthquake-proof, 8, 60, 93–100, **107**, 109

continental drift, theory of, 7, 41–46, 49–52, 57, 63–64, 68–69, 71–72, 74, 76, 78, 108
convergence, **80**, 81–82
Copernicus, Nicolaus, 20

Darwin, Charles, 7, 35
divergence, **80**, 81

earth, **12**
　belief in flat, 8, **10**, 11–13
　circumference, 13–16, **16**
　early beliefs about formation, 20, 32–33, 69
　heliocentric model of, 20, 108
　layers of, 25, 27, **28**, 29, 53, 69, 74, 76, 78–79, 85
earthquakes

in California, 60, **94**, 97, 99–100
in Chile, 39
in China, 14, 36, 98–99
early beliefs about, 13–16, 36
forecasting probability of, 95–97
improving safety in earthquake-prone areas, 8, 60, 93–100, **107**, 109
in Italy, 18, 36
in Japan, 5, 14, 19, 36, 64–65, 84, 97–98
measuring and studying, 36–41, 58–61, 64–66, 77, 83
in Mexico, 5–6, 84, 97
in Nepal, 84
why they happen, 83–88
Eratosthenes, 13
Erta Ale lava lake, 5, **104**, 105
Ethiopia, 5, 105–106

fault lines, 18, 83, 85, 91
fossils, study of, 6, 13, 15, 20–25, 30, 32, 41, 44

geology, 7–8, 15, 24–27, 35, 41, 44, 49, 68–69, 108
birth of modern, 32–35
geomythology, 15
glaciers, 44–45
Gutenberg, Beno, 38–39, 60–63, 66

Hawaii, 11, 14, **88**, 89
Hess, Hary, 46, **62**, 63–64, 74
HMS *Challenger*, **34**, 35
Holmes, Arthur, **67**, 68–69, 71–72
hot spots, 89
Hutton, James, 7, 32–35

Iceland, 14, 53, 76
igneous rock, 30, 33
Indonesia, **9**
Io, 100–103, **101**

lava, 5–6, 18, **31**, 85–86, **104**, 105
Leonardo da Vinci, 6, 20–24, **22**, 27, 29
Liu, Lijun, 91

maps/atlases, 6, 41–42, 73–74, 77

McKenzie, Dan, 52–53, 55, 58, 76–77, 88, 108
Mercalli scale, 38–39
metamorphic rock, 33–34
Mid-Atlantic Ridge, 74–76, **75**
Milne, John, 36–38, 84
moment magnitude scale, 41, 83–84
Morgan, Jason, 53, **54**, 55–58, 69, 77, 88, 108
Mori Tower, 98–99
Moro, Anton, 30, 32–33

oceans, exploration and study of, 7, 35–36, 41, 44, 49, 55, 63, 66, 73
  mapping ocean floor, 73–76
  seafloor spreading, 46, 57, 63, 73–74, 76, 81
original horizontality, principle of, 25–27, 29
Ortelius, Abraham, 42, 44, 50

paleontology, 7, 20, 41, 49
Pangaea, 43, 45, **47**, **70**, 71, 78
Pangaea Ultima, 106

plate tectonics
  and "critical mass," 56–57, 69
  discovery of, 6, 8, 29, 53, 55–58, 63–64, 69, 71–89
  explaining, 78–79
  future of the earth and, 103–109
  influence of discovery, 93–103
  major players in discovery of, 49–69
  number of plates, 78–79
  plate movement, 79–83, **80**, 91, 93, 106
  in solar system, 8, 100–103
  technological advancements and, 72–73, 90–91, 109
  when/how it began, 88
Pompeii, 6, **17**, 17–19, 27

radioactive decay, 68
Renaissance, science during the, 6, 8, 20–27, 29
Richter, Charles, 38–39, 58–61, **59**, 65–66, 95

Richter scale (Gutenberg-
    Richter scale), 38–41,
    58–61
  diagram, **40**
Ring of Fire, 65, 84–85
rock, types of, 30, 33–34
Rossi-Forel intensity scale,
    38

sedimentary rock, 30, 33
seismographs and
    seisometers, 36–38, **37**,
    41, 60, 66, 83, 96
seismology, 8, 41, 56, 60, 64,
    66, 83
Shanghai Tower, 98–99
sonar, 63, 73–74
Steno, Nicolaus, 24–27,
    29–30
Strabo, 16–17
subduction, 66, 81–82, **82**
supercomputing models,
    90–91, 109
superposition, law of, 27, 29

Tharp, Marie, 74–76
transform boundaries, 83
Transamerica Pyramid, 92,
    99–100
tsunami, **9**, **19**, 85

in Japan, 5–6, 19–20,
    84–85, 98

uniformitarianism, principle
    of, 33

Volcán de Fuego, **26**
volcanoes, **4**, 5–6, 23, **26**, 27,
    30, **31**, **87**
  creation of new islands,
      103–105
  early beliefs about, 11,
      14–17
  elsewhere in solar system,
      8, 100–103
  in Japan, 36
  Pompeii and Mount
      Vesuvius, 17–18, 27
  volcanic rock, 30, 33
  why they happen, 85–88

Wadati, Kiyoo, 64–66
Wegener, Alfred, 7–8,
    41–46, **48**, 49–52, 57–58,
    63, 68–69, 71, 74, 76, 78,
    108
Wilson, John Tuzo, 89

About the Author

Fiona Young-Brown has written a number of books, including *Eleanor Roosevelt: First Lady*, *Nuclear Fusion and Fission*, and *The Universe to Scale: Similarities and Differences in Objects in Our Solar System*. She also enjoys writing about food, travel, and great apes. Originally from England, Young-Brown now lives in Kentucky with her husband and two dogs.